Cambridge Tracts in Mathematics and Mathematical Physics

J. G. LEATHEM, M.A.

E. T. WHITTAKER, M.A., F.R.S.

No. 11

The Fundamental Theorems of the Differential Calculus

THE FUNDAMENTAL THEOREMS

OF THE

DIFFERENTIAL CALCULUS

by

W. H. YOUNG, Sc.D., F.R.S.

Formerly Fellow of Peterhouse, Cambridge
Lecturer on Higher Analysis at the University of Liverpool

CAMBRIDGE:
at the University Press
1910

CAMBRIDGE
UNIVERSITY PRESS

University Printing House, Cambridge CB2 8BS, United Kingdom

Cambridge University Press is part of the University of Cambridge.

It furthers the University's mission by disseminating knowledge in the pursuit of education, learning and research at the highest international levels of excellence.

www.cambridge.org
Information on this title: www.cambridge.org/9781107493629

First published 1910
Re-issued 2015

A catalogue record for this publication is available from the British Library

ISBN 978-1-107-49362-9 Paperback

Cambridge University Press has no responsibility for the persistence or accuracy of URLs for external or third-party internet websites referred to in this publication, and does not guarantee that any content on such websites is, or will remain, accurate or appropriate.

PREFACE

IN this Tract rigidity of proof and novelty of treatment have been aimed at rather than simplicity of presentation, though this has never been lightly sacrificed. The Differential Calculus is concerned with those continuous functions that possess differential coefficients and with these differential coefficients themselves. As a differential coefficient is not necessarily a continuous function, the subject merges naturally into the wider one of the Theory of Functions of one or more Real Variables, and cannot, therefore, be completely mastered without some knowledge of the Theory of Sets of Points. No more knowledge of the language or concepts of this theory will however here be required than a serious mathematical student may now be supposed to have gained before completing his Degree course, and, with this exception, the present account of the fundamental theorems of the Differential Calculus will, it is hoped, be found to be complete in itself. For the rest a brief account is given in Appendix III of the definitions and results from the Theory of Sets of Points actually employed in the Tract.

The theory of those functions that correspond to the differential coefficient at a point at which this latter does not exist, does not fall naturally within our scope. Some of the remarkable properties of these interesting functions, with other generalities, are, however, stated without proof in Appendix I.

The theory of Maxima and Minima has been barely alluded to, and the complex variable has been rigorously excluded. Apart from other considerations the space at our disposal has here been decisive. For the same reason it has been impossible to give more than a few isolated examples. Appendix II consists of references to some of the existing literature, where such examples may be found.

It is perhaps well to warn the English reader in conclusion that such initial difficulties as he may feel are likely to be due, in part at least, to a lack of familiarity with the modern formulation of the concept of an irrational number.

W. H. YOUNG.

LA NONETTE DE LA FORÊT,
GENEVA, SWITZERLAND.
November, 1909.

CONTENTS

XII. ON THE REVERSIBILITY OF THE ORDER OF PARTIAL DIFFERENTIATION.

XIII. POWER SERIES.

XIV. TAYLOR'S THEOREM.

APPENDICES.

I. PRELIMINARY NOTIONS.

1. Independent variables. Interval. Right and left.

In the present tract we shall deal with one or more real independent variables x, each of which assumes all values in some interval, which may be *closed* $(a \leqslant x \leqslant b)$ or *completely open* $(a < x < b)$ or *half-open*. The values of x may be supposed represented in any of the usual ways on a straight line, which will be laid horizontal and so that, as x increases, the representative point moves to the right. This straight line will be imagined closed at each end by a point, the point $+ \infty$ on the right and $- \infty$ on the left, and when we speak of the interval (a, b), the left-hand end-point may, unless the contrary is stated, be $- \infty$, and the right-hand end-point $+ \infty$.

If there are two or more independent variables, these will be represented on rectangular axes, and the ensemble of them (x, y), or $(x_1, x_2, ..., x_n)$, will be represented by a point in the plane or higher space. The correlative of an interval will then be a rectangle, or n-dimensional parallelepiped, and may be closed or completely or partially open. In what immediately follows x will, for brevity, be used equally for a single variable and for the ensemble of several variables, and the word *interval* will be used with the understanding that in higher space the proper interpretation is to be put upon it.

2. Function of one or more variables. Finite. Bounded.

If to each x there is an unique value f, this is said to be a *function* of x, and we write $f(x)$ for it. It is found convenient to include the two distinct infinite numbers $+ \infty$ and $- \infty$ as among the values which the most general kind of function may assume[a].

If $f(x)$ has at each point x a finite value, it is said to be a *finite function*. If there is some finite number greater (less) than any value of the function, $f(x)$ is said to be *bounded above (below)*, and if bounded both above and below, $f(x)$ is a *bounded function*.

Ex. 1. Let $f(0) = 0$, $f(x) = \dfrac{1}{x}$, $(x \neq 0)$. Then f is a finite function, unbounded both above and below.

1

II. LIMITS.

3. Upper and lower bound. Upper and lower limit.

Unique limit. The least (greatest) number which is not less (greater) than any value of a function is called its *upper (lower) bound*. It is therefore the same thing to say a function is unbounded above (below) or that the upper (lower) bound is $+ \infty$ ($- \infty$).

If we take a sequence of intervals* each inside the preceding,

$$d_1, d_2, ..., d_n, ...,$$

and having one and only one common internal point a, the upper (lower) bounds of $f(x)$ for values of x other than a in these successive intervals will not increase (decrease), so that they form a monotone descending (ascending) series

$$U_1 \geqslant U_2 \geqslant ... \geqslant U_n \geqslant ...,$$
$$L_1 \leqslant L_2 \leqslant ... \leqslant L_n \leqslant ....$$

The lower bound ϕ of the former and the upper bound ψ of the latter series are therefore unaltered if we omit any finite number of its constituents, or if we interpolate the upper (lower) bound taken with respect to any intermediate interval. Hence it is easily perceived that ϕ and ψ do not depend on the particular sequence of intervals chosen, provided only the same point a is in each case the sole and only common internal point. Thus ϕ and ψ depend only on the point a and the function f, and are called respectively *the upper and lower limits of f at the point a.* As we vary a, ϕ and ψ become functions $\phi(x)$ and $\psi(x)$, and are called *the associated upper and lower limiting functions* [β] *of f(x)*.

It may sometimes be convenient in considering the behaviour of $f(x)$ in the neighbourhood of the point a to omit other values of x besides the value a, which is always omitted. The values of x which are retained must form a set S with a as limiting point, and should in each case be expressly defined. We then get, by a precisely similar method, *upper and lower limits at the point a with respect to the set S.* Each such limit is said to be *a limit* of $f(x)$ at the point a, and there will always be *a plurality of limits*, except when the upper and lower limits at a coincide, in which case $f(x)$ is said to have *an unique limit* at the point a, and its value is the common value of the upper and lower limits. When there is a single independent variable x, it is important to consider the case in which the set S consists of all

---

* See Appendix III.

points of an interval on one side only of the point a. We thus get upper and lower and possibly intermediate *limits on the right and on the left* respectively of the point a, and, when these coincide, we get *an unique limit on the right or on the left*.

4. Plurality of Limits. The concept of a plurality of limits can only be clearly grasped by reference to the elementary facts of the Theory of Sets of Points. As the function $y = f(x)$ is supposed to be any whatever, the mode of distribution of the set of points G_d on the y-axis, which represents the values of the function at all points except a of some interval d containing a on the x-axis, is any whatever. Certain of these points may be repeated an infinite number of times, since the function may assume the same value over and over again. If this is the case, we add such points to the first derived set* of G_d. The set H_d so formed is still a closed set, since the addition of points of a set to its first derived set introduces no new limiting points. In particular the set H_d includes the point Q_d which represents the upper bound of $f(x)$ in the interval d considered.

If we now let the interval d shrink up to the point a, as in the preceding article, the successive closed sets H_d so obtained lie each inside the preceding sets, and therefore, by Cantor's Theorem of Deduction*, determine a closed set H, consisting of all their common points, which is easily seen to be the same, however the interval d shrinks up to the point a, and includes the unique limiting point Q of the points Q_d. *This closed set H of values of y is said to constitute the set of limits of the function $f(x)$ at the point a.* Since the set H certainly includes the point Q, the set of limits includes the upper (and similarly the lower) limit as defined in the preceding article, and this whether these limits are taken with respect to the continuum or any other set.

On the other hand, since every set of points contains at least one sequence* having as unique limiting point any required limiting point of the original set, it follows that, if l be any limit of $f(x)$ at $x = a$, there is a sequence $x_1, x_2, \ldots x_n, \ldots$ having a as unique limiting point, such that, passing along this sequence, $f(x)$ has l as unique limit. This important property proves at the same time, in conjunction with the preceding paragraph, that the limits[v] as defined in the present article are the same as in the preceding article.

We use the notation

$$\underset{x=a}{\mathrm{Llt}}\, f(x)$$

* See Appendix III.

to denote the set of all the limits of $f(x)$ at the point a, while, if it is known that there is an unique limit, we write $\underset{x=a}{\mathrm{Lt}}\, f(x)$ for that limit. Thus the equation

$$y = \underset{x=a}{\mathrm{Lt}}\, f(x)$$

must be understood to mean both that $f(x)$ has an unique limit at $x = a$, and also that that limit is y.

5.   Double and repeated limits.   Any repeated limit is a double limit.

If there are two independent variables x and y, the limits at $(a,\ b)$, obtained in the manner explained in §§ 3, 4, with the two-dimensional interpretation given in § 1, are called *double limits* of $f(x,\ y)$ at $(a,\ b)$.

Similarly if there are n variables, the corresponding limits are called n-ple limits at the point.

If $f(x,\ y)$ is a function of two independent variables x and y, it becomes a function of x alone when we keep y constant and has a corresponding set of simple limits

$$\underset{x=a}{\mathrm{Llt}}\, f(x,\ y).$$

If there is only one such limit for each value of y, this limit defines a function of y, and has as such a set of limits for $y = b$; these are called *the repeated limits* of $f(x,\ y)$ first with respect to x and then with respect to y, and written

$$\underset{y=b}{\mathrm{Llt}}\, \underset{x=a}{\mathrm{Lt}}\, f(x,\ y).$$

Similarly, if there is an unique limit when y is kept constant,

$$\underset{x=a}{\mathrm{Llt}}\, \underset{y=b}{\mathrm{Lt}}\, f(x,\ y)$$

denotes *the repeated limits* of $f(x,\ y)$, first with respect to y and then with respect to x.

Sometimes it is desirable to consider only such double limits as result from values of $f(x,\ y)$ in a neighbourhood of the point $(a,\ b)$ other than points on the axial cross through $(a,\ b)$, that is on $x = a$ and on $y = b$. Such a neighbourhood is called *a non-axial neighbourhood*. In particular we have the following theorem.

THEOREM.   *Any repeated limit is a double limit, taken with respect to a non-axial neighbourhood of the point considered.*

For by the definition of a repeated limit, say u, whatever sequence of values different from a, say $x_1,\ x_2 \ldots,\ \ldots x_n,\ \ldots$ be taken, having a as unique limit, $f(x_n, y)$ has, for fixed y different from b, an unique limit, say $v(y)$, when n is indefinitely increased, and the quantities $v(y)$ have at $y = b$ the repeated limit u in question as one of their limits ; that is

$$v(y) = \underset{n=\infty}{\text{Lt}} f(x_n, y), \quad (y \neq b),$$

$$u = \text{one of the} \underset{y=b}{\text{Llt}} v(y).$$

Represent the values of these functions on a straight line, as in § 4,

$$f(x_n, y) \text{ by the point } P_{n,y},$$

$$v(y) \text{ by the point } Q_y,$$

$$u \text{ by the point } Q.$$

Then, unless the points $P_{n,y}$, for fixed y, all coincide with the exception of a finite number of them, Q_y is the unique limiting point of the points $P_{n,y}$, while in the excluded case, it is the repeated point itself. In either case, denoting as in § 4 these points $P_{n,y}$, which as n and y vary, lie in a neighbourhood d of the point a, by the set G_d, each point Q_y is a point of the closed set H_d, consisting of the first derived set of G_d together with its repeated points, if any.

Again Q is either a repeated point of the set H_d or one of its limiting points, and therefore in any case it is for all neighbourhoods d a point of the set H_d, since that set is closed. Thus Q represents by § 4 a double limit of $f(x, y)$ at the point (a, b). Moreover, since the points (x_n, y) all lie off the axial cross, this double limit is taken with respect to a non-axial neighbourhood of the point (a, b).

It is an immediate consequence that *if $f(x, y)$ has an unique double limit, and the simple limit*

$$\underset{h=0}{\text{Lt}} f(a+h, b+k)$$

is unique for all values of k in a certain neighbourhood of $k = 0$, then the repeated limit

$$\underset{k=0}{\text{Lt}} \underset{h=0}{\text{Lt}} f(a+h, b+k)$$

exists and is equal to the double limit.

It may evidently happen that, even when $\underset{x=a}{\text{Llt}} f(x, y)$ do not all coincide, the upper and lower limits, and therefore all intermediate limits, have an unique limit, which is the same for each, as y approaches b. If we agree to call this an unique repeated limit, it is evident that the above reasoning holds, with the small modification that the values $x_1, x_2, \ldots$ are not independent of y, but form for each value of y a sequence with the same property as before. Thus such an unique repeated limit is also a double limit.

It need hardly be remarked that the existence of an unique limit of $f(x, y)$ for each fixed value of y, does not, of course, involve that of an unique limit when y varies with x, which would be concomitant to the existence of an unique double limit.

The most general definition of a repeated limit, when neither of the simple limits involved is unique, and the corresponding theorem, which still holds, are not required in the present connexion [23].

III. CONTINUITY AND SEMI-CONTINUITY.

6. Upper and lower semi-continuity. If the upper limit at a is $\leqslant$ the value of f at that point, that is if (§ 3)

$$\phi(a) \leqslant f(a),$$

$f(x)$ is said to be *upper semi-continuous* at the point a. If on the other hand

$$f(a) \leqslant \psi(a)$$

$f(x)$ is said to be *lower semi-continuous* at a. If at every point of an interval, open or closed, $f(x)$ is upper (lower) semi-continuous it is said to be *an upper (lower) semi-continuous function*. In particular it is easily proved that $\phi(x)$ is an upper and $\psi(x)$ a lower semi-continuous function [δ].

Ex. 2. Let $f(x) = 0$, when x is zero or irrational, and

$$f(x) = \frac{1}{q}, \text{ when } x = \frac{p}{q} \text{ is rational, } (0 < x \leqslant 1),$$

p and q being integers prime to one another.

Then $f(x)$ is an upper semi-continuous function, and its $\psi(x) = 0$, $\phi(x) = 0$. By changing the sign of f we get a lower semi-continuous function.

It follows immediately from the definitions that *the points, if any, at which an upper (lower) semi-continuous function assumes values $\geqslant k$ ($\leqslant k$) form a closed set.*

Another important property is the following:—*An upper (lower) semi-continuous function assumes in every closed interval its upper (lower) bound in that interval.* For let $k_1 < k_2 < \ldots < k_n < \ldots$ be an ascending sequence of numbers having the upper bound U of an upper semi-continuous function $f(x)$ for upper bound. Then if G_n denote the closed set of points at which $f(x) \geqslant k_n$, there is, by Cantor's Theorem of Deduction, at least one point a common to all the successive sets G_n. Hence $k_n \leqslant f(a) \leqslant U$ for all integers n. But U is the upper bound of the k's, therefore $f(a) = U$.

Hence it easily follows[e] that *an upper (lower) semi-continuous function which is finite in (a, b) is bounded in some interval inside (a, b).*

THEOREM. *A monotone decreasing sequence of functions $f_1 \geqslant f_2 \geqslant \ldots$ which are upper semi-continuous at a point P has for limit a function f which is also upper semi-continuous at P.*

For at a point where $f = +\infty$, it is, of course, upper semi-continuous, we have therefore only to prove the theorem at a point where

$$f(P) < A,$$

A being a finite quantity.

Since $f(P)$ is the limit of $f_n(P)$, we can determine m so that

$$f_m(P) < A,$$

and, since f_m is upper semi-continuous, we can find an interval d, containing P as internal point, such that *throughout it*

$$f_m(x) < A.$$

Since the sequence of functions is monotone decreasing, it follows that, for all values of $n \geqslant m$,

$$f_n(x) < A,$$

so that, throughout the interval d,

$$f(x) \leqslant A.$$

Since A was any quantity greater than $f(P)$, this shews that f is upper semi-continuous at P, which proves the theorem.

In the same way we have the corresponding theorem :—*A monotone increasing sequence of functions $f_1 \leqslant f_2 \leqslant \ldots$ which are lower semi-continuous at P, has for limit a function f which is also lower semi-continuous at P.*

7. Continuity. If a function $f(x)$ is both upper and lower semi-continuous at the point a, that is, if it has an unique limit at a whose value is the same as $f(a)$, the function is said to be *continuous* at the point a. If it is continuous at every point of an interval, open or closed, it is said to be *a continuous function* in that interval.

We already know therefore that a continuous function has the following properties :—

The points, if any, at which a continuous function assumes values $\leqslant k$ and those at which it is $\geqslant k$, and therefore those at which it is $= k$, all form closed sets.

A continuous function assumes its upper and lower bounds in every closed interval.

It has also the following important property, which, as will be seen in the sequel, is not confined to continuous functions, and is shared by all differential coefficients, whether or no they are continuous :—

A continuous function assumes all values between its upper and lower bounds.

In fact since a continuum cannot be divided into two closed sets, the two closed sets of points at which respectively $f \geqslant k$ and $f \leqslant k$ (k being any value between the upper and lower bounds of f), must have a common point, at which accordingly $f = k$.

It follows from the second of the above properties that *a finite continuous function is bounded.* The word *continuous* is therefore often used as synonymous with *finite and continuous*, and this usage will in the present tract be adopted, except where the contrary is stated. It is in this sense that the word *continuous* must be understood in the following alternative definition which may be called the ϵ-definition of continuity.

A function is said to be continuous at a point if, given any positive quantity ϵ, however small, we can find a closed interval with that point as internal point, such that the difference between the upper and lower bounds of the function—the so-called oscillation of the function—in that interval is less than ϵ.

The theorem of § 6 gives us a standard test for continuity, viz. *a function which is at the same time the limit of a monotone descending and of a monotone ascending sequence of continuous functions is a continuous function.* It may be remarked that this condition is not only sufficient but also necessary[7].

The set of limits at a point of discontinuity of a function which is continuous throughout an open interval ending at that point is of the simplest character, namely a closed interval. To prove this we merely have to remark that if k is any value less than $\phi(a)$ and greater than $\psi(a)$, $f(x)$ assumes values both $> k$ and $< k$, and therefore assumes the value k at a point of the open interval. Thus k is certainly one of the limits, since it is a repeated value (§ 4). Therefore the set of limits consists of every value from $\phi(a)$ to $\psi(a)$ inclusive.

8. Pointwise discontinuous function. A function which, without being necessarily continuous, has in *every* interval a point of continuity is called a *pointwise discontinuous function.* In other words, the points of continuity of such a function are *everywhere dense*, without necessarily filling up any interval. It will now be proved that a finite *semi-continuous function is pointwise discontinuous* [5].

For simplicity of wording, we shall suppose $f(x)$ to be finite and lower semi-continuous, so that inside any chosen interval we can choose an interval in which $f(x)$ is bounded (§ 6) and

$$f(x) \leqslant \psi(x) \leqslant \phi(x) \dots\dots\dots\dots\dots\dots(1).$$

We remark first, that, whatever be the nature of a bounded function $f(x)$, the points at which

$$\phi(x) - f(x) \geqslant k \dots\dots\dots\dots\dots\dots\dots(2)$$

can in no case fill up an interval. For, if a were an internal point of such an interval, we should have

$$\text{upper } \operatorname*{Lt}_{x=a} \phi(x) \geqslant \text{upper } \operatorname*{Lt}_{x=a} f(x) + k \geqslant \phi(a) + k,$$

which is not true, since, by § 6, $\phi(x)$ is upper semi-continuous and is bounded, since $f(x)$ is bounded.

In our case $\phi(x) - f(x)$ is the excess of an upper over a lower semi-continuous function and is therefore upper semi-continuous. Hence the points at which (2) holds, form a closed set, which is therefore, by what precedes, *dense nowhere*. There must therefore be an interval throughout which

$$\phi(x) - f(x) < k.$$

Repeating this process in this interval with $\tfrac{1}{2}k$ for k, and so on, we arrive at a point, internal to all the successive intervals, at which

$$\phi(x) - f(x) < \frac{k}{2^n}$$

for every value of the integer n. Therefore, since the left-hand side is by (1) not negative, it must, at this point, be zero. Hence, using (1) again, at this point,

$$f(x) = \psi(x) = \phi(x),$$

which proves the function to be continuous there, and therefore to possess a point of continuity in every interval.

We have seen that the limit of a monotone sequence of continuous functions is a semi-continuous function, it follows therefore from the above that it is a pointwise discontinuous function. Baire has proved[5], [8] that *the limit of a sequence of continuous functions is always pointwise discontinuous,* and not only so but that it *is pointwise discontinuous with respect to every perfect set*; that is to say, approaching a certain point a of any perfect set by means of points of that set in any manner, we shall get for $f(x)$ the unique limit $f(a)$.

The following example shews that, even when the sequence of continuous functions is monotone, the limiting function need not be continuous.

Ex. 3.   Let $f_n(x)$ have at all the points $\dfrac{p}{q}$, where $q < n+2$, the value $\dfrac{1}{q}$, and between these points be linear, so that the locus $y = f_n(x)$ is a broken line.   These functions are continuous, and they form a monotone descending sequence.   Their limiting function is the function $f(x)$ of Ex. 2, § 6.

IV.   DIFFERENTIATION.

9. The Incrementary Ratio. Derivates. Differential coefficient. Second and higher differential coefficients; these are repeated limits.

If $f(x)$ is a (finite) continuous function of a single real variable x throughout a closed interval (a, b), *the incrementary ratio*

$$m(x, y) \equiv \frac{f(x) - f(y)}{x - y}$$

is a function of the ensemble (x, y) defined at all points of a certain closed square except on the diagonal $x = y$, and is continuous at every point at which it is defined.   It has, like a proper continuous function, the property of *assuming every value between its upper and lower bounds*, as may be proved without difficulty[15].

The limits with respect to y of $m(x, y)$ at any point on the diagonal are called the *derivates* of $f(x)$ at the point x; they are called *right or left-hand derivates*, according as $y > x$ or $y < x$, and, in particular, the upper and lower limits on the two sides are called *the upper and lower derivates* on the two sides[v].

If the derivates all coincide, their common value is called *the differential coefficient of $f(x)$*.   In other words $f(x)$ has at the point x a differential coefficient provided

$$m(x, x + h) \equiv \frac{f(x + h) - f(x)}{h}$$

has as h approaches zero in any manner an unique limit, and this limit is the value of the differential coefficient; if this is true when h is positive (negative) the value is that of the *right-hand (left-hand) differential coefficient* *.

Here we are expressly assuming that $f(x)$ is finite and continuous. At a point at which $f(x)$ is not finite, or is discontinuous, we shall say that a differential coefficient does not exist.   It may also not exist at a point at which $f(x)$ is continuous.   It should, however, be remarked

---

* Notice that by the differential coefficient at the left (right) hand end-point of the interval considered we mean the right (left) hand differential coefficient.

that there is nothing in the definition to prevent $f(x)$ having a differential coefficient at a point x at which it is continuous, even when there is no interval containing the point *throughout* which $f(x)$ is continuous.

It follows from the definition that a differential coefficient is not necessarily finite, it may have the value $+\infty$ or $-\infty$ at particular points. At such a point the differential coefficient, which is denoted by $f'(x)$, or $\dfrac{dy}{dx}$, has not itself a differential coefficient. Moreover, by Baire's theorem, quoted in the last article, $f'(x)$ is a pointwise discontinuous, not necessarily continuous, function of x, if it exists at every point of an interval.

If $f'(x)$, distinguished in this connexion as *the first differential coefficient*, exists throughout an interval, and is continuous at the point x, and has a differential coefficient there, that differential coefficient is called *the second differential coefficient* of $f(x)$, and is denoted by $f''(x)$, or $\dfrac{d^2f}{dx^2}$. In like manner *the higher differential coefficients* of $f(x)$, if they exist, are defined successively, each being the differential coefficient of its predecessor.

Thus the assumption that an nth differential coefficient, or *differential coefficient of the n-th order*, $f^{(n)}(x)$, or $\dfrac{d^nf}{dx^n}$, exists at a point, carries with it necessarily the existence, the finiteness and the continuity of all preceding differential coefficients at the point, and their existence and finiteness in some neighbourhood of the point. It does not imply the finiteness of the nth differential coefficient itself at the point, nor its existence in the neighbourhood, still less its continuity at the point.

We may also define the higher differential coefficients as repeated limits, e.g. the second differential coefficient as the repeated limit, if unique [θ],

$$\underset{k=0}{\mathrm{Lt}}\ \underset{h=0}{\mathrm{Lt}}\ \frac{f(x+h+k)-f(x+k)-f(x+h)+f(x)}{hk}.$$

10. Differentiation of a function of a function. Differentiation of the sum, difference, product and quotient of two functions. Leibniz's rule. It at once follows that, *if u is a function of x having a differential coefficient at a certain point, and x is a function of t having a differential coefficient at the* *corresponding point, then u is a function of t having a differential coefficient at that point, and it is given by the formula*

$$\frac{du}{dt} = \frac{du}{dx} \cdot \frac{dx}{dt}.$$

For by hypothesis to each value of t in the given interval there corresponds a value of x in the interval (a, b) and to this x a value of u. Hence u may be regarded as a function of t; say,

$$u = f(x) = F(t) \dots\dots\dots\dots\dots(1),$$
$$x = g(t) \dots\dots\dots\dots\dots\dots(2),$$
$$x + h = g(t + \tau) \dots\dots\dots\dots\dots(3).$$

Now however τ approaches the limit zero, $g(t+\tau)$ approaches the unique limit $g(t)$, since g is continuous. Hence, by (2) and (3) h has the unique limit zero. Using the simultaneous approach of h and τ to zero in the identity

$$\frac{F(t+\tau) + F(t)}{\tau} = \frac{f(x+h) - f(x)}{h} \cdot \frac{g(t+\tau) - g(t)}{\tau} \dots\dots(4)$$

the right-hand side has, under the specified conditions, the unique limit $\frac{du}{dx} \cdot \frac{dx}{dt}$, so that the left-hand side has also the same unique limit, which proves the theorem.

By repeated application of the above rule it follows that, if $f(x)$ has an nth differential coefficient with respect to x, and x has one with respect to t, then f is a function of t having an nth differential coefficient with respect to t, which is found by a simple rule.

By a still more immediate application of the theory of limits, we obtain the theorem that *if two functions u and v have each a differential coefficient with respect to x at a particular point, so have their sum, difference, product and quotient, and these differential coefficients are respectively*

$$\frac{du}{dx} + \frac{dv}{dx}; \quad \frac{du}{dx} - \frac{dv}{dx}; \quad v\frac{du}{dx} + u\frac{dv}{dx}; \quad -\frac{1}{v^2}\left(v\frac{du}{dx} - u\frac{dv}{dx}\right).$$

These are, for the rest, special cases of the formula of § 21.

Repeated application of these rules gives us the nth differential coefficient of the compound function when u and v have each nth differential coefficients. In particular we get *Leibniz's rule for the n-th differential coefficient of a product*, which may be written symbolically

$$\frac{d^n}{dx^n}(uv) = \left(u\frac{dv}{dx} + v\frac{du}{dx}\right),$$

where the right-hand side is to be expanded by the Binomial Theorem, and then $\frac{d^r v}{dx^r}$ substituted for $\left(\frac{dv}{dx}\right)^r$ and $\frac{d^r u}{dx^r}$ for $\left(\frac{du}{dx}\right)^r$.

V. INDETERMINATE FORMS.

11. The indeterminate forms $\frac{0}{0}$, $\frac{\infty}{\infty}$, first rule. A differential coefficient is the first example encountered by the student of what is called an *indeterminate form*.

The Theory of Indeterminate Forms has usually been based on the Theorem of the Mean [1], given below (§ 15); it can be developed independently, and perhaps still more simply, as the following shews.

THEOREM. *If as x approaches the value a, f(x) and F(x) have both the unique limit zero, or + ∞ or − ∞, then the limits of*

$$f(x)/F(x) \quad\dots\dots\dots(1)$$

lie between* the upper and lower limits of

$$f'(x)/F'(x)\dots\dots\dots(2),$$

provided

A. *a is not a limiting point of common infinities of $f'(x)$ and $F'(x)$;*

B. *a is not a limiting point of zeros of $F'(x)$ unless these zeros are also zeros of $f'(x)$†; and*

C. *$F(x)$ is monotone.*

CASE 1. Let the unique limit of $f(x)$ and of $F(x)$ be zero.

Assume for definiteness that $F(x)$ never decreases as x increases, so that $F(x)$ is positive and $F'(x)$ is never negative in a sufficiently small neighbourhood $(a < x)$. It will suffice to prove that the upper limit of (1) is ⩽ the upper limit of (2); for the same argument, *mutatis mutandis*, will prove that the lower limit of (1) is ⩾ the lower limit of (2).

We may suppose that there is a finite quantity L greater than the upper limit of (2), for, if not, that upper limit would be $+\infty$, and what is required to be proved is obvious.

We then have

$$\frac{f'(x)}{F'(x)} - L < 0,$$

* We shall say that x lies *between* a and b when x is a point of the closed interval (a, b).

† It will be seen from what follows later in the Tract, that C. is included in B. should a not be a limiting point of zeros of $F'(x)$.

at all points at which $f'(x)$ and $F'(x)$ are not simultaneously zero in a sufficiently small neighbourhood $(a < x)$. Moreover at such common zeros,

$$f'(x) - LF'(x) = 0.$$

Noting further that, if the point a is a limiting point of infinities of $F'(x)$, none of which can accordingly by A. be infinities of $f'(x)$, L is certainly positive, we have, in all cases, and for all points in such a neighbourhood,

$$f'(x) - LF'(x) \leqslant 0 \quad\dots\dots\dots\dots\dots\dots(3).$$

Now since $f'(x)$ and $F'(x)$ are never infinite together, the left-hand side of (3) is the differential coefficient of $f(x) - LF(x)$, whence it follows that $f(x) - LF(x)$ never increases. But zero is the unique limit of $f(x) - LF(x)$ at the point a, therefore this function is never positive in the neighbourhood considered. Hence, as $F(x)$ is positive,

$$\frac{f(x)}{F(x)} - L \leqslant 0$$

throughout the neighbourhood. Therefore the upper limit of (1) as x approaches a is less than or equal to L. But L was any finite quantity greater than the upper bound of (2). Hence the upper limit of (1) is less than or equal to the upper limit of (2), as was to be shewn.

CASE 2. $f(x)$ and $F(x)$ each have an infinite limit.

We may clearly without loss of generality suppose the two infinite limits to be $+\infty$, since otherwise we need only change the sign of one or both functions. Similar considerations to those used above enable us to assume that $F(x)$ is always positive and $F'(x)$ never positive, throughout the chosen interval, and to assume that there is a finite quantity L greater than the upper limit of (2).

Then, as before, remembering that $F'(x)$ is now $\leqslant 0$,

$$f'(x) - LF'(x) \geqslant 0,$$

the convention being made, as before, that since $F'(x) \leqslant 0$, (2) has the opposite sign to $f'(x)$, so that, when L is finite, $f'(x)$ cannot be negative at a point where $F'(x)$ is zero.

Hence $f(x) - LF(x)$

never decreases as x increases from a, and has therefore its lower bound as its unique limit at $x = a$.

Now suppose, if possible, that (1) had a limit greater than L, say $L + 2e$. Then

$$f(x) - (L + e)F(x), \text{ that is } f(x) - LF(x) - eF(x)$$

is positive for some sequence of values of x with a as limiting point.

But $F(x)$, and therefore $eF(x)$, has $+\infty$ as unique limit at $x=a$, therefore the same is true of $f(x)-LF(x)$ for the sequence in question. Hence, by the above, the lower bound of $F(x)$ is $+\infty$, which is absurd. This proves the theorem, which, it will be noticed, is somewhat wider in its scope than that usually given.

COR. *Under the same provisos as in the theorem, if* (2) *has an unique limit, so has* (1). It must of course not be supposed that, if (1) has an unique limit, (2) has one.

We have tacitly assumed a to be at a finite distance, and the approach to be in the direction of x decreasing. It is clear that the argument is perfectly general, and the approach may be in the other direction. Also a may be either $+\infty$ or $-\infty$, in which cases the approach will be in the appropriate direction.

12. Second rule for $\dfrac{0}{0}$. It should be noticed that in the preceding theorem no assumption is made as to the existence of $f'(x)$ and $F'(x)$ at the point a, nor indeed do we assume that they have each an unique limit as x approaches a. In the following corollary we assume the existence of certain differential coefficients at the point a, but not the existence, even implicitly, of the last of these differential coefficients in the neighbourhood of the point a.

COR. *If $f(x)$ and $F(x)$ vanish when $x=a$ and at that point have respectively their first $(r-1)$ and $(s-1)$ differential coefficients zero, while their r-th and s-th differential coefficients respectively are finite, and not zero, then*

$$f(x)/F(x) \quad \dotfill (1)$$

has an unique limit as x approaches a, and the value of the limit is

$$0, \quad f^{(r)}(a)/F^{(r)}(a), \quad or \quad \pm\infty,$$

according as $r>s$, $r=s$, or $r<s$; moreover the sign of the infinity in the last case is $+$ or $-$ according as $f^{(r)}(a)$ and $F^{(s)}(a)$ have the same or opposite signs.

Since $f^{(n)}(x)$ exists at $x=a$, all the preceding differential coefficients exist in a closed neighbourhood of $x=a$. Hence

$$\operatorname*{Lt}_{h=0} \frac{f(a+h)}{h^r} = \operatorname*{Lt}_{h=0} \frac{f'(a+h)}{rh^{r-1}} = \dots = \operatorname*{Lt}_{h=0} \frac{f^{(r-1)}(a+h)}{r!\,h},$$

but since $f^{(r-1)}(a)=0$,

$$\frac{f^{(r-1)}(a+h)}{h} = \frac{f^{(r-1)}(a+h)-f^{(r-1)}(a)}{h},$$

so that, by the definition of $f^{(r)}(a)$, the last expression has the limit $f^{(r)}(a)$. Hence

$$\underset{h=0}{\text{Lt}} \frac{f(a+h)}{h^r} = \frac{1}{r!} f^{(r)}(a).$$

Similarly,

$$\underset{h=0}{\text{Lt}} \frac{F(a+h)}{h^s} = \frac{1}{s!} F^{(s)}(a).$$

But

$$\frac{f(a+h)}{F(a+h)} = h^{r-s} \frac{f(a+h)/h^r}{F(a+h)/h^s},$$

whence the theorem at once follows, since h^{r-s} and the numerator and denominator of the fraction multiplying it have unique limits.

13. The Expansion Theorem

$$f(a+h) = f(a) + hf'(a) + \ldots + \frac{1}{n!} h^n \{f^{(n)}(a) + \epsilon\}.$$

The property of possessing a differential coefficient of higher order than the first, say of the nth order, can in part[κ] be expressed in a form precisely analogous to what may be called the ϵ-definition of a differential coefficient. This form is of importance both for its own sake and because it suggests the corresponding theorem, much less easily proved, concerning functions of more than one variable.

THEOREM. *If $f(x)$ possesses an n-th differential coefficient at $x = a$, then*

$$f(a+h) = f(a) + hf'(a) + \tfrac{1}{2}h^2 f''(a) + \ldots$$

$$+ \frac{1}{(n-1)!} h^{n-1} f^{(n-1)}(a) + \frac{1}{n!} h^n \{f^{(n)}(a) + \epsilon\}$$

where ϵ has zero as unique limit when h approaches zero in any manner whatever.

When $n = 1$, the theorem is, as already remarked, identical with the definition.

Put, in the general case,

$$F(h) = f(a+h) - f(a) - hf'(a) - \ldots \frac{h^{n-1}}{(n-1)!} f^{(n-1)}(a).$$

Further, put

$$G(h) = h^n.$$

Then both $F(h)$ and $G(h)$ are finite and continuous functions of h in an interval containing $h = 0$, and $G(h)$ is monotone. Also, by the hypothesis made,

$$F(h), \ F'(h), \ \ldots, \ F^{(n-1)}(h),$$
$$G(h), \ G'(h), \ \ldots, \ G^{(n-1)}(h),$$

all are zero at the origin; hence, by the theory of indeterminate forms,

$$\underset{h=0}{\text{Lt}}\frac{F(h)}{G(h)} = \underset{h=0}{\text{Lt}}\frac{F^{(n-1)}(h)}{G^{(n-1)}(h)} = \underset{h=0}{\text{Lt}}\frac{f^{(n-1)}(a+h)-f^{(n-1)}(a)}{n!\,h} = \frac{f^{(n)}(a)}{n!},$$

by definition. Hence the required result follows.

It should be noticed that the form of the result stated in the theorem is, like most statements in ϵ-language, lacking in generality, it assumes the finiteness of $f^{(n)}(a)$, unless, as we are at liberty to do, we interpret the statement of the theorem suitably in the case when $f^{(n)}(a)$ is infinite, namely as equivalent to the statement in the language of limits just obtained.

COR. *The reasoning remains unaltered if, instead of h^n, we substitute any function of h, all of whose differential coefficients up to the $(n-1)th$ inclusive vanish at $h=0$, and having $n!$ as n-th differential coefficient, provided only the function be a monotone function of h, and the same is true of all its differential coefficients concerned.*

Thus we might take the function

$$\frac{n!}{g^{(n)}(a)}\left\{ g(a+h)-g(a)-\dots-\frac{h^{n-1}}{(n-1)!}g^{(n-1)}(a)\right\}$$

provided $g^{(n)}(a)$ is not zero.

VI. MAXIMA AND MINIMA.

14. Maxima and minima of a function of a single variable. The theorem just proved gives us at once the following criterion for a maximum or minimum:—*If $f(x)$ possesses an r-th differential coefficient at the point $x=a$ which is not zero, and if all the preceding differential coefficients are zero there, then if r is even, $f(x)$ is a maximum or minimum at the point $x=a$, according as the sign of the r-th differential coefficient is negative or positive. If r is odd, there is neither a maximum nor a minimum.*

It should be noticed that this theorem is true whether or not the rth differential coefficient is finite or infinite, as follows from the corresponding remark as to the theorem of the last article.

It should also be noticed that we have not assumed the existence of the rth differential coefficient except at the point $x=a$, still less its continuity there or elsewhere.

VII. THE THEOREM OF THE MEAN.

15. The Theorem of the Mean. A differential coefficient assumes all values between its upper and lower bounds. A differential coefficient is one of the limits of differential coefficients in the neighbourhood. Before proceeding further we must prove the following fundamental theorem known as the Theorem of the Mean.

THEOREM. *If $f(x)$ has$^{(\lambda)}$ a differential coefficient $f'(x)$ at every point of the completely open interval (a, b), and is continuous also at a and b, there is a point x of the completely open interval (a, b) at which this differential coefficient is equal to the incrementary ratio $m(a, b)$, that is*

$$f'(x) = \frac{f(b) - f(a)}{b - a} = m(a, b).$$

CASE 1 (*Rolle's Theorem*). Let

$$f(b) = f(a) = 0.$$

The theorem is evident in the trivial case when $f(x) = 0$ at every point of (a, b). If this is not the case, $f(x)$ has a positive upper bound or a negative lower bound, or both, and, being a continuous function, assumes such an extreme value at a point x of the closed interval (a, b); and this point cannot be one of the end-points since the function is zero there. At such a point x the numerators in both the incrementary ratios ✶

$$m(x + h, x) = \frac{f(x + h) - f(x)}{h}$$

and

$$m(x - h, x) = \frac{f(x - h) - f(x)}{-h},$$

where not zero, have the same sign for every value of h, not necessarily the same in the two expressions, less than a certain quantity.

Hence, making h approach zero in both incrementary ratios, $f'(x)$ is both $\leqslant 0$ and $\geqslant 0$, so that

$$f'(x) = 0,$$

which proves the theorem.

CASE 2. Let

$$f(b) \neq f(a).$$

Let z denote the ordinate-distance of a point on the locus $y = f(x)$ from the chord AB joining the two points whose abscissae are a and b.

Then z only differs by a constant from $f(x) - mx$, where m is written for $m(a, b)$. Thus z has a differential coefficient at every point of the closed interval (a, b), and is zero at the end-points. Hence, by Rolle's Theorem,

$$\frac{d}{dx}(f(x) - mx) = 0$$

at some point x of the completely open interval (a, b), that is

$$f'(x) - m = 0,$$

which proves the theorem.

COR. 1. *The upper and lower bounds of the differential coefficient in any interval, open or closed, are the same as those of the incrementary ratio.*

COR. 2. *The differential coefficient like the incrementary ratio assumes every value between its upper and lower bounds in any closed interval at points internal to that interval.*

COR. 3. *The differential coefficient $f'(x)$ is one of the limits of the differential coefficients in each neighbourhood of the point x.*

For at some point ξ of the completely open interval $(x, x + h)$

$$f'(\xi) = m(x + h, x).$$

As we let $h \geqslant 0$ (or $h \leqslant 0$) move up to zero continuously, ξ will move up to x in a certain determinate manner, not necessarily continuously. Moving ξ in this manner, $f'(\xi)$ has the same unique limit as $m(x + h, x)$, that is $f'(x)$.

Ex. $\qquad\qquad f(0) = 0, \qquad (x \leqslant 0),$

$$f(x) = x^2 \sin \frac{\pi}{x}, \qquad (0 < x).$$

Here $\qquad f'(x) = 2x \sin \frac{\pi}{x} - \pi \cos \frac{\pi}{x}, \qquad (0 < x),$

so that any number from $-\pi$ to π, both inclusive, is a limit of the differential coefficients on the right of the origin.

But at the origin $f'(0) = 0$, so that the differential coefficient at the origin is *one among* the limits of differential coefficients in the neighbourhood.

COR. 4. *If $f'(x) = 0$ at every point of an interval, $f(x)$ is constant throughout that interval.*

COR. 5. *If a differential coefficient exist throughout an interval, the points at which it is finite must be dense everywhere.*

The remaining points have been shewn to form a set of content zero [25].

It must not be supposed that a differential coefficient cannot be zero at points dense everywhere in an interval throughout which the differential

2—2

coefficient exists, without the original function being a constant (μ). Köpcke was the first to construct such a non-constant function, a so-called *every-where-oscillating function* [17].

VIII. PARTIAL DIFFERENTIATION AND DIFFERENTIALS.

16. Partial differential coefficients of various orders. If f is a function of two or more variables $(x, y, ...)$, it becomes a function of a single variable x when we keep all the remaining variables constant, and as such it may have a differential coefficient, called the *partial differential coefficient with respect to* x and usually denoted by f_x or $\frac{\partial f}{\partial x}$, to distinguish it from the total differential coefficient $\frac{df}{dx}$ obtained by making the remaining variables arbitrary functions of x and differentiating by the rule of § 10 ; in English writings this symbolism is not always strictly adhered to, and $\frac{df}{dx}$ is sometimes used for the partial differential coefficient when there is no danger of ambiguity.

Thus

$$\frac{\partial f}{\partial x} \equiv \underset{h=0}{\text{Lt}} \frac{f(x+h, y, ...) - f(x, y, ...)}{h}$$

and is a function of $(x, y, ...)$. It may therefore have a partial differential coefficient with respect to each variable, viz.

$$\frac{\partial}{\partial x} \cdot \frac{\partial f}{\partial x} = \frac{\partial^2 f}{\partial x^2} = f_{xx}, \quad \frac{\partial}{\partial y} \cdot \frac{\partial f}{\partial x} = \frac{\partial^2 f}{\partial y \partial x} = f_{yx}, \text{ etc.}$$

These are called *partial differential coefficients of* $f(x)$ *of the second order* and define in like manner those of the third order, and so on.

It is an immediate result of these definitions that f_{xy} and f_{yx} are repeated limits of the double incrementary ratio

$$m(a, b\; ; \; a+h, b+k)$$

$$\equiv \frac{f(a+h, b+k) - f(a+h, b) - f(a, b+k) + f(a, b)}{hk}.$$

By repeated application of the Theorem of the Mean it follows that if f_{xy} exists at every point of a closed rectangle $(a, b\; ; \; a+h, b+k)$ there is an internal point (x_1, y_1) of the rectangle such that

$$m(a, b\; ; \; a+h, b+k) = f_{xy}(x_1, y_1).$$

Hence it follows that f_{xy} and f_{yx}, if they exist throughout the rectangle, assume all values between their upper and lower bounds [ν]— since, as is easily proved, the double incrementary ratio does so— and also that f_{xy}, f_{yx} and $m(x, y; x', y')$ have the same upper and lower bounds [ρ].

A knowledge of these partial differential coefficients, which then take the place of the successive differential coefficients of a function of a single variable, no longer gives us the equivalent information. The importance in the case of a function of a single variable of the differential coefficients consists in the fact that, from the very definition of a differential coefficient, the theorem of § 13 holds for $n = 1$, and hence, also, as we have seen, for all values of n. No corresponding theorem, even in the case of $n = 1$, follows from the mere existence of the partial differential coefficients. In the theory of functions of two or more variables the proper correlative of the differential coefficient, in the case of a function of a single variable, is not the ensemble of the partial differential coefficients of the first order, but what is called the *differential*. We proceed therefore to give a brief account of its theory, confining our attention in the first instance to two variables.

17. Differential. *A function $f(x, y)$ of two real variables is said to have a total first differential at the point (a, b), if*

A. *The partial differential coefficients*
$$f_a \equiv \frac{\partial f(a, b)}{\partial a} \quad and \quad f_b \equiv \frac{\partial f(a, b)}{\partial b}$$
both exist and are finite at the point (a, b).

B. *At all points $(a + h, b + k)$ of a closed neighbourhood of the point (a, b) the function can be expressed as follows*
$$f(a + h, b + k) = f(a, b) + hf_a + kf_b + he + ke' \quad\ldots\ldots(1),$$
where e and e' have each the unique double limit zero when h and k approach zero in any manner whatever.

Note 1. For brevity the word *total* is usually omitted.

Note 2. $hf_a + kf_b$ is often called *the first differential* of $f(x, y)$ and denoted by df. It should be noticed that this name is only properly applied when an equation of the form (1) holds.

18. Sufficient conditions for the existence of a first differential.

LEMMA. *In order that $f(x, y)$ may have a first differential, it is*

*sufficient, but not necessary, beside the obvious condition that f_a and f_b
should both exist and be finite, that one of the two incrementary ratios*

$$m_{a+h}\,(b,\ b+k) \equiv \frac{f(a+h,\ b+k)-f(a+h,\ b)}{k},$$

$$m_{b+k}(a,\ a+h) \equiv \frac{f(a+h,\ b+k)-f(a,\ b+k)}{h},$$

*should have an unique double limit when h and k approach zero in any
manner whatever.*

This is an immediate consequence of the identities

$$f(a+h,\ b+k)-f(a,\ b) = hm_b\,(a,\ a+h)+km_{a+h}\,(b,\ b+k)$$
$$= hm_{b+k}(a,\ a+h)+km_a\,(b,\ b+k).$$

Cor. *If either f_x or f_y is a continuous function of the ensemble
$(x,\ y)$ at the point $(a,\ b)$, while the other exists at that point, then $f(x,y)$
has a first differential at $(a,\ b)$.*

For in this case, by the Theorem of the Mean, the above two
incrementary ratios have the same limit as

$$f_y\,(a+h,\ b+\theta k)\ \text{and}\ f_x\,(a+\theta h,\ b+k),\qquad (0<\theta<1),$$

viz., f_b and f_a respectively.

19. The Fundamental Theorem of Differentials.

Theorem. *If $\partial f/\partial x$ and $\partial f/\partial y$ each have differentials of the first
order at the point $(x,\ y)$, then*

$$\frac{\partial}{\partial x}\frac{\partial f}{\partial y} = \frac{\partial}{\partial y}\frac{\partial f}{\partial x},$$

*and their common value is also the unique double limit of the double
incrementary ratio*

$$m\,(x,\ y;\ x+h,\ y+k) \equiv \frac{f(x+h,\ y+k)-f(x+h,\ y)-f(x,\ y+k)+f(x,\ y)}{hk}$$

as h and k approach zero in any manner whatever.

Since f_x has a differential,

$$f_x\,(x+h,\ y+k)-f_x(x,\ y) = h\,(f_{xx}+e)+k\,(f_{yx}+e')\quad \ldots\ldots(1),$$

where e and e' have zero as unique double limit as h and k approach
their limit zero. Now

$f_x\,(x+h,\ y+k)-f_x(x,\ y)$

$$=\frac{f_x\,(x+h,\ y+k)-f_x(x+h,\ y)}{k}\,k +\frac{f_x\,(x+h,\ y)-f_x(x,\ y)}{h}\,h$$

$$=\frac{f_x\,(x+h,\ y+k)-f_x\,(x+h,\ y)}{k}\,k +(f_{xx}+e'')\,h\ \ldots\ldots(2),$$

where e'' vanishes with h, however k varies.

Equating the right-hand sides of (1) and (2) by dividing by k, we get

$$\frac{f_x(x+h, y+k) - f_x(x+h, y)}{k} = \frac{h(e-e'')}{k} + f_{yx} + e' \ldots\ldots(3).$$

Hence, however h and k proceed to their limit zero, provided h/k does not become infinite, that is k/h does not become zero, the right-hand side of (3) has the limit f_{yx}, so that the same is true of the left-hand side of (3). In the same way we prove, by considering f_y and interchanging x and y, that $\dfrac{f_y(x+h, y+k) - f_y(x, y+k)}{h}$ has the unique limit f_{xy} for all modes of approach of (h, k) to $(0, 0)$, provided h/k has not zero for limit.

Now, writing $m_y(x, x+h) \equiv \dfrac{f(x+h, y) - f(x, y)}{h}$,

we have by the Theorem of the Mean

$$m(x, y; x+h, y+k) = \frac{m_{y+k}(x, x+h) - m_y(x, x+h)}{k} = \frac{d}{dy} m_{y+\theta k}(x, x+h),$$

where $0 < \theta < 1.$

Hence $m(x, y; x+h, y+k) = \dfrac{f_y(x+h, y+\theta k) - f_y(x, y+\theta k)}{h} \ldots(4).$

Now, if (h, k) moves towards $(0, 0)$ in such a way that h/k has not zero as limit, $h/\theta k$ will not have zero for a limit; and therefore, by what has been proved, the right-hand side of (4) will have the unique double limit f_{xy}, and therefore the same is true of $m(x, y; x+h, y+k)$.

Similarly, if (h, k) moves towards $(0, 0)$ in such a way that k/h has not zero for a limit, m will have f_{yx} for unique double limit.

Since we can choose a mode of approach of (h, k) to $(0, 0)$ for which neither h/k nor k/h has zero for a limit, it follows that $f_{xy} = f_{yx}$. Hence for every mode of approach $m(x, y; x+h, y+k)$ has an unique limit whose value is $f_{xy} = f_{yx}$; this proves the theorem.

COR. *If throughout a closed neighbourhood of a point (a, b) the $(n-1)th$ partial differential coefficients of $f(x, y)$ all exist and are independent of the order of differentiation, and have first differentials at (a, b), then the order of differentiation is indifferent in the n-th differential coefficients at the point (a, b).*

For a formal proof of this corollary the reader is referred to the original memoir [20].

20. Second and higher differentials. The higher differentials are defined, like the higher differential coefficients, successively, and, like the differential coefficients, must be understood only to exist at a point when all the preceding differentials exist not only at that point, but also in a closed neighbourhood of that point. Another analogy between the definitions consists in the fact that whereas the first differential, like the first differential coefficient, is defined by means of a process involving the function $f(x, y)$ itself, the higher differentials, like the higher differential coefficients, are defined by means of the same process involving not the function itself directly but the preceding differential.

DEFINITION OF THE SECOND DIFFERENTIAL d^2f. *If the first differential df, in which we regard h and k as arbitrary constants, has a first differential at the point (a, b), this differential is called the second differential of f at the point (a, b), and is denoted by d^2f.*

This is clearly equivalent to the following:

If f_x and f_y exist in a closed neighbourhood of the point (a, b), and have first differentials at the point (a, b), *then provided f has a first differential in the neighbourhood of* (a, b),

$$hdf_x + kdf_y$$

is called the second differential of f, or, say,

$$d^2f = hdf_x + kdf_y = h\left(hf_{xx} + kf_{yx}\right) + k\left(hf_{xy} + kf_{yy}\right).$$

It should be noticed that we are following closely the analogy with the theory in the case of one independent variable. We define, in fact, $f'(x)$ as the limit of

$$\frac{f(x+h)-f(x)}{h},$$

that is, virtually by means of the equation

$$f(x+h)-f(x) = h[f'(x)+e],$$

but we do not define $f''(x)$ by a similar expansion statement, not, for example, as the limit of

$$\frac{f(x+h)-f(x)-hf'(x)}{\tfrac{1}{2}h^2},$$

but as the differential coefficient of $f'(x)$.

It is worth remarking that the mere fact that f_x and f_y have first differentials at the point (a, b), necessitating as it does the continuity of f_x and f_y with respect to the ensemble (x, y) at the point (a, b), involves the existence of df there, but not, of course, necessarily in the neighbourhood of that point.

By the Fundamental Theorem of Differentials (§ 19) it now follows that *when d^2f exists,*

$$f_{ab} = f_{ba},$$

so that we may write the second differential in the following form

$$d^2 f = h^2 f_{aa} + 2hk f_{ab} + k^2 f_{bb},$$

or, symbolically, $\quad d^2 f = \left(h \dfrac{\partial}{\partial a} + k \dfrac{\partial}{\partial b} \right)^2 f.$

DEFINITION OF THE nTH DIFFERENTIAL $d^n f$. *If the $(n-1)$th differential, in which we regard h and k as arbitrary constants, has a first differential at the point (a, b), this differential is called the n-th differential of f at the point (a, b), and is denoted by $d^n f$.*

Here, as before, we get

$$d^n f = h^n d f_{a^n} + n h^{n-1} k \, d f_{a^{n-1} b} + \cdots$$

$$= \sum_{r=0}^{r=n} \binom{n}{r} h^{n-r} k^r \frac{d^n f}{da^{n-r} db^r} = \left(\frac{d^n f}{da^n}, \cdots \right) \! \backslash \! \left(h, \ k \right)^n,$$

or, symbolically,

$$= \left(h \frac{\partial}{\partial a} + k \frac{\partial}{\partial b} \right)^n f,$$

and we shew that the order of differentiation is indifferent to the partial differential coefficients of the nth order at the point (a, b), using for this purpose the Corollary of § 19.

It follows at once from the definitions that, *if f has an n-th differential at (a, b), then df has an $(n-1)$th differential at (a, b), and that f itself has all the differentials up to the $(n-1)$th inclusive, not only at (a, b) but also in a suitable neighbourhood of the point (a, b).*

21. Successive differentiation of a function of two or more functions. It follows from the definition of a differential that, *if u is a function of x and y, and x and y are functions of t, and if, for any particular value of t, u possesses a differential with respect to (x, y), and x and y possess differential coefficients with respect to t, then u is a function of t which possesses a differential coefficient there, whose value is given by the equation*

$$\frac{du}{dt} = \frac{\partial u}{\partial x} \cdot \frac{dx}{dt} + \frac{\partial u}{\partial y} \cdot \frac{dy}{dt},$$

or, if there are more subsidiary functions (see below § 24),

$$\frac{du}{dt} = \sum_i \frac{\partial u}{\partial x_i} \cdot \frac{dx_i}{dt}.$$

Moreover, successive application of the Fundamental Theorem of Differentials gives us the following important result :

If $f(x, y)$ possesses an n-th differential at the point considered, then we may obtain the n-th differential coefficient with respect to t by *repeated application of this rule, and may, at each stage of the process, regard the order of partial differentiation with respect to x and y as indifferent up to the highest differential coefficients which occur.*

We may remark that, in the application of this rule, it is unnecessary to assume that the highest differential coefficients which occur are continuous.

The above rule, which includes as special cases the rules for the differentiation of a sum, difference, product or quotient of two functions, is formally unaltered when we come to deal with functions of more variables possessing differentials, except that the total differential coefficients are replaced by differentials. We have in fact the following theorem:

If a function f of two or more variables x_i possess a differential with respect to them, and each x_i is a function of certain other variables t_n, and possesses a differential at a certain point $(T_1, T_2, ...)$, then f is a function of the variables t_n and possesses a differential at the point in question, given by

$$df = \sum_i \frac{\partial f}{\partial x_i} dx_i.$$

Similar remarks to those made above about the repeated application of the rule apply here. In particular Leibniz's formula for the successive differential coefficients of a product is valid for differentials, with the same change of the symbol $\frac{d}{dt}$ into d.

22. Theorem of the Mean for two variables. Corresponding to the Theorem of the Mean for one variable, we have the following Theorem of the Mean for two variables:

If $f(a, b) = 0$ and $f(A, B) = 0$, while $f(x, y)$ has a first differential throughout an area containing the whole stretch from (a, b) to (A, B), then there is a point (x, y) internal to that stretch at which

$$(x - a)f_x + (y - b)f_y = 0.$$

Put $x = a + (A - a)t, \qquad y = b + (B - b)t$;

then, by the Theorem of the Mean for a single variable,

$$\frac{df}{dt} = 0$$

for some value of t internal to the interval $(0, 1)$, which corresponds therefore to a point (x, y) internal to the stretch from (a, b) to (A, B).

But, since $f(x, y)$ has a differential throughout the area considered,

$$0 = \frac{df}{dt} = (A - a)f_x + (B - b)f_y = \frac{A - a}{x - a}[(x - a)f_x + (y - b)f_y],$$

since
$$\frac{x-a}{A-a} = \frac{y-b}{B-b},$$
which proves the theorem.

23. The Expansion Theorem

$$f(a+h,\ b+k) = f(a,\ b) + df(a,\ b) + \ldots + \frac{1}{n!}\,(d^n f + E_n).$$

We are now in a position to state and prove the expansion theorem for two variables which corresponds to that given in § 13 for a single variable. We begin by repeating the following statement:

If $f(x, y)$ has a first differential at the point (a, b), we may expand $f(x, y)$ in the neighbourhood of (a, b) in the following form:

$$f(a+h,\ b+k) = f(a,\ b) + df(a,\ b) + he_1 + ke_2,$$

where e_1 and e_2 have each the unique double limit zero when h and k approach zero in any manner whatever.

This is, in fact, only another form of writing the equation used in the definition of the first differential.

We next prove a second special case of the expansion theorem we have in view, viz.:

If $f(x, y)$ has a first differential at the point (a, b), and $\partial f/\partial x$, $\partial f/\partial y$ have first differentials at the same point, then

$$f(a+h,\ b+k)$$
$$= f(a,\ b) + df(a,\ b) + \tfrac{1}{2}\,(h^2 f_{aa} + 2hk f_{ab} + k^2 f_{bb}) + \tfrac{1}{2}\,(h^2 e_1 + 2hk e_2 + k^2 e_3),$$

where e_1, e_2 and e_3 each have the unique double limit zero when h and k approach zero in any manner whatever.

For the conditions of § 19 are satisfied, so that $m(a+h, b+k;\ a, b)$ has the unique limit f_{ab} or f_{ba}. Hence

$$f(a+h,\ b+k) - f(a+h,\ b) - f(a,\ b+k) + f(a,\ b) = hk\,(f_{ab} + e_2),$$

where e_2 has the unique double limit zero.

But, since f_{aa} and f_{bb} exist and are finite, we have, by one-dimensional theory (§ 13),

$$f(a+h,\ b) - f(a,\ b) = hf_a + \frac{h^2}{2!}\,(f_{aa} + e_1),$$

$$f(a,\ b+k) - f(a,\ b) = kf_b + \frac{k^2}{2!}\,(f_{bb} + e_3),$$

where e_1 vanishes with h, independently of k, and e_3 vanishes with k, independently of h.

Adding the three last equations, the theorem follows.

Note. The condition that $f(x, y)$ has a first differential at the point is, of course, included in the further conditions of the enunciation.

COR. *If $f(x, y)$ has a second differential at the point (a, b),*
$$f(a + h, b + k) = f(a, b) + df(a, b) + \tfrac{1}{2}d^2f(a, b) + \tfrac{1}{2}(h^2e_1 + 2hke_2 + k^2e_3).$$

We now proceed to prove the general theorem of which that of § 13 is a particular case.

THEOREM. *If $f(x, y)$ has an $(n-1)$th differential at the point (a, b), where $n \geqslant 3$, while the $(n-1)$th differential coefficients exist and are independent of the order of differentiation in a closed neighbourhood of the point (a, b), and have first differentials at (a, b), then the "n-th expansion theorem" holds, that is to say*
$$f(a + h, b + k)$$
$$= f(a, b) + df(a, b) + \tfrac{1}{2}d^2f(a, b) + \dots + \frac{1}{n!}\left(\frac{d^nf}{da^n}, \dots \right)\!\!\left(h, k\right)^n + \frac{1}{n!}E_n,$$

where
$$E_n = \sum_{r=0}^{r=n} \binom{n}{r} e_r h^{n-r} k^r = (e_0, e_1, \dots, e_n)(h, k)^n,$$

and the quantities e_r all have zero as unique double limit when h and k approach zero in any manner whatever.

Let $l_0, l_1, \dots, l_n$ be any functions of a, b, h, and k, such that identically
$$0 = f(a + h, b + k) - f(a, b) - df(a, b) - \tfrac{1}{2}d^2f(a, b) - \dots$$
$$- \frac{1}{(n-1)!}d^{n-1}f(a, b) - \frac{1}{n!}(l_0, l_1, \dots, l_n)(h, k)^n \dots(1).$$

In the right-hand side of this identity change h into $(x - a)$ and k into $(y - b)$, excepting only in $l_0, l_1, \dots, l_n$, which are left unaltered, and denote the result by $g(x, y)$, so that
$$g(x, y) = f(x, y) - f(a, b) - \left\{\frac{\partial f}{\partial a}(x - a) + \frac{\partial f}{\partial b}(y - b)\right\}$$
$$- \frac{1}{2!}\left(\frac{\partial^2 f}{\partial a^2}, \dots \right)\!\!\left(x - a, y - b\right)^2 - \dots$$
$$- \frac{1}{(n-1)!}\left(\frac{\partial^{n-1} f}{\partial a^{n-1}}, \dots \right)\!\!\left(x - a, y - b\right)^{n-1}$$
$$- \frac{1}{n!}(l_0, l_1, \dots, l_n)(x - a, y - b)^n \dots\dots\dots(2).$$

Then $\qquad g(a, b) = 0, \qquad g(a + h, b + k) = 0.$

Also, since $f(x, y)$ has a first differential in the neighbourhood of the point (a, b), $g(x, y)$ has the same property, so that, by § 22, there is a point (x, y) internal to the stretch from (a, b) to $(a + h, b + k)$, such that
$$(x - a)g_x + (y - b)g_y = 0 \dots\dots\dots\dots\dots(3).$$

Now, since

$$\frac{1}{m!}\frac{\partial}{\partial x}(A_0, A_1, \ldots, A_m \big\rangle\!\big\langle x-a, y-b)^m$$

$$\equiv \frac{1}{(m-1)!}(A_0, A_1, \ldots, A_{m-1}\big\rangle\!\big\langle x-a, y-b)^{m-1},$$

and

$$\frac{\partial^m f}{\partial a^{m-r+1}\partial b^{r-1}} = \frac{\partial^{m-1}f_a}{\partial a^{m-r}\partial b^{r-1}},$$

we get, by differentiating (2),

$$g_x = f_x - f_a - \left[(x-a)\frac{\partial f_a}{\partial a} + (y-b)\frac{\partial f_a}{\partial b}\right] - \cdots$$

$$- \frac{1}{(n-2)!}\left(\frac{\partial^{n-2}f_a}{\partial a^{n-2}}, \ldots \big\rangle\!\big\langle x-a, \ y-b\right)^{n-2}$$

$$- \frac{1}{(n-1)!}(l_0, l_1, \ldots, l_{n-1}\big\rangle\!\big\langle x-a, y-b)^{n-1}.$$

Thus, *if the $(n-1)$th expansion theorem holds for f_x,*

$$g_x = \frac{1}{(n-1)!}\left(\frac{\partial^{n-1}f_a}{\partial a^{n-1}}, \ldots \big\rangle\!\big\langle x-a, y-b\right)^{n-1}$$

$$+ \frac{1}{(n-1)!}(e_0', e_1', \ldots, e_{n-1}'\big\rangle\!\big\langle x-a, y-b)^{n-1}$$

$$- \frac{1}{(n-1)!}(l_0, l_1, \ldots, l_{n-1}\big\rangle\!\big\langle x-a, y-b)^{n-1}$$

$$= \frac{1}{n!}\frac{\partial}{\partial x}\left(\frac{\partial^n f}{\partial a^n} - l_0, \ldots, \frac{\partial^n f}{\partial b^n} - l_n\big\rangle\!\big\langle x-a, y-b\right)^n$$

$$+ \frac{1}{(n-1)!}(e_0', e_1', \ldots, e_{n-1}'\big\rangle\!\big\langle x-a, y-b)^{n-1},$$

where the e's have zero as limit, when h and k, and therefore $(x-a)$ and $(y-b)$, have zero as limit.

A similar expression holds for g_y, *provided the $(n-1)$th expansion theorem holds for f_y,* this expression involving quantities $e_1'', e_2'', \ldots, e_n''$, which again have zero as double limit.

Thus, by (3),

$$0 = (x-a)g_x + (y-b)g_y = \frac{1}{(n-1)!}\left(\frac{\partial^n f}{\partial a^n} - l_0, \ldots, \frac{\partial^n f}{\partial b^n} - l_n\big\rangle\!\big\langle x-a, y-b\right)^n$$

$$+ \frac{1}{(n-1)!}(e_0, e_1, \ldots, e_n\big\rangle\!\big\langle x-a, y-b)^n \ldots(4),$$

where

$$n e_r = (n-r)e_r' + r e_r''$$

for all values of r, e_n and e_0'' having each the value zero; thus the e's have zero as unique double limit when h and k approach zero in any manner whatever.

But $$\frac{x-a}{h} = \frac{y-b}{k},$$

so that, from (4),

$$(l_0, l_1, \ldots, l_n \mathbin{\text{$\rangle$}} h, k)^n = \Big(\frac{d^n f}{da^n}, \ldots \mathbin{\text{$\rangle$}} h, k\Big)^n + E_n,$$

and substituting this in (1), the required expansion follows. This proves the theorem when $n = 3$. For we are given that f_{xx}, f_{xy}, f_{yy} have first differentials at the point, so that, as proved on p. 27, f_x and f_y have the second expansion property required in the above proof.

If $n > 3$, the theorem will be proved if we are able to shew that f_x and f_y possess the $(n-1)$th expansion property.

Proceeding on the same lines as before, we reduce the task of shewing that this is the case to that of proving that f_{xx}, f_{xy} and f_{yy} possess the $(n-2)$th expansion property. We know, in fact, that f_x and f_y have $(n-2)$th differentials at (a, b), since f has an $(n-1)$th differential at that point. Therefore, since $n - 2 > 1, f_x$ and f_y have second differentials at (a, b), and therefore, by the definition, first differentials in a closed neighbourhood of (a, b), which is required for the application of the lemma. Also we know that f_{xx}, f_{xy} and f_{yy} have $(n-3)$th differentials, since f has an $(n-1)$th differential at (a, b). Hence the argument already used applies *mutatis mutandis*.

Proceeding thus successively we reduce the problem to shewing that the $(n-2)$th partial differential coefficients have the second expansion property, which follows from p. 27, since we are given that their differential coefficients have first differentials at (a, b).

Thus in any case the theorem is demonstrated.

COR. *If f possesses an n-th differential at the point (a, b), then*

$$f(a+h, b+k) = f(a, b) + df + \frac{1}{2!} d^2 f + \ldots + \frac{1}{n!} d^n f + \frac{1}{n!} E_n.$$

24. Differentials of a function of more than two variables. We have hitherto tacitly assumed more than once that the theory of the differentials of functions of two variables is typical of that in the case of more variables. At the present stage it is advisable to examine what modifications are necessary in §§ 19—23 to make them apply to functions of n variables where $n > 2$. We have then n, instead of two, partial differential coefficients, $f_{x_1}, f_{x_2}, \ldots$ which, as in § 17, are assumed to be finite, and to appear linearly in the equation

$$f(x_1, x_2, \ldots, x_n) = f(a_1, a_2, \ldots, a_n) + \sum_r (x_r - a_r)(f_{a_r} + e_r),$$

the quantities e_1, e_2, ..., e_n having zero as unique n-ple limit when the h's approach zero in any manner whatever. As in § 18 we get a set of sufficient conditions for the existence of a differential at a certain point, demanding that one of the partial incrementary ratios should have an unique n-ple limit, one an unique $(n-1)$-ple limit, and so on, one an unique double limit and the last an unique simple limit, viz. the corresponding partial differential coefficient. This gives the corollary :—*If at the point* $(a_1, a_2, ..., a_n)$ *all the partial differential coefficients exist, all but one are continuous with respect to* (x_1, x_2), *all but one of those remaining with respect to* (x_1, x_2, x_3), *and so on, finally the last one is continuous with respect to* $(x_1, x_2, ..., x_n)$, *then* $f(x_1, x_2, ..., x_n)$ *has a first differential at the point* $(x_1, x_2, ..., x_n)$. The fundamental theorem of § 19 applies as it stands, since all but two of the independent variables may be kept constant during the discussion. Hence the definitions of the second and higher differentials and the remaining investigations of §§ 20 and 21 apply, as well as the statement and proof of the Theorem of the Mean for n variables (§ 22), the equation involved being now

$$\sum_r (x_r - a_r) f_{x_r} = 0.$$

As regards § 23, the reasoning on p. 27 is no longer sufficient; the proof on pp. 28, 29 however applies when $n > 2$, and also when $n = 2$, provided we add to the assumptions the express condition that $f(x_1, x_2, ..., x_n)$ has a first differential *in the neighbourhood* of the point $(a_1, a_2, ..., a_n)$, in which case we have the condition of the corollary on p. 28. With this gloss the expansion theorem of § 23 certainly holds for any number of variables. The conditions for a maximum or minimum are therefore as stated in § 25.

IX. MAXIMA AND MINIMA FOR MORE THAN ONE VARIABLE.

25. Maxima and minima of a function of two or more variables. Want of space forbids us to enter into a long discussion of this subject, but it is worth while noting that the result just obtained gives us the following rule, analogous to that already stated for one dimension :—*If f possesses a non-zero n-th differential at the point (a, b) and if all preceding differentials vanish there, then, unless n is odd, f is neither a maximum nor a minimum, while, if n is even, f is a minimum if the n-th differential be essentially positive, and a maximum if it be essentially negative.*

Here two things should be noted. No assumption is made as to the continuity of the nth differential coefficients at the point (a, b), indeed we do not even assume their existence in the neighbourhood of the point (a, b). In the second place, it is to be understood that the nth differential is not to be zero for any values of h and k other than zero.

X. EXTENSIONS OF THE THEOREM OF THE MEAN.

26. The Remainder Form of Taylor's Theorem; Lagrange's form.

The fact that we can pass from the definition of a differential coefficient to an equation of the form given in § 15, suggests the possibility of extending the Theorem of the Mean so as to involve the nth differential coefficient. We propose here to state and prove the following theorem :

THEOREM. *If $f^{(n-1)}(a)$ exists and is finite, and $f^{(n)}(x)$ exists throughout the completely open interval $(a, a+h)$, then*

$$f(a + h - 0) = f(a) + hf'(a) + \dots$$
$$+ \frac{1}{(n-1)!} h^{n-1} f^{(n-1)}(a) + \frac{1}{n!} f^{(n)}(a + \theta h), \quad (0 < \theta < 1),$$

where $f(a + h - 0)$ is any one of the limits of the function $f(x)$ as x approaches the value $a + h$.

[NOTE. Here $f(x)$, $f'(x)$, $\dots$, $f^{(n-2)}(x)$ are necessarily continuous in the half-open interval $(a \leqslant x < a + h)$, but we do not assume the continuity at the point $x = a$ of $f^{(n-1)}(x)$, nor the existence of f or any of its differential coefficients at $x = a + h$, nor do we assume the finiteness of $f^{(n)}(a)$ in the open interval.

The case of this theorem when the interval considered is completely closed and all the functions up to $f^{(n-1)}$ at least are continuous throughout the whole closed interval is often, for a reason that will appear later, referred to as Lagrange's Remainder Form of Taylor's Theorem.]

We require first of all a slight modification of the Theorem of the Mean, which is easily proved, viz. the following :

If $f(a + 0)$ and $f(b - 0)$ denote respectively any one of the limits of $f(x)$ to the right of a and any one to the left of b, then, provided
(1) $f(x)$ is continuous and finite in the open interval, $a < x < b$;
(2) $f(x)$ has a differential coefficient at every point of the open interval,
then there is a point x of the open interval such that

$$f(b - 0) - f(a + 0) = (b - a) f'(x).$$

If f is itself a differential coefficient of another function, and exists at the point a without necessarily being continuous there, then one of the values of $f(a+0)$ is $f(a)$ itself, and we may write

$$f(b-0)-f(a)=(b-a)f'(x).$$

Now supposing the theorem proved for the integers $1, 2, \ldots,$ $(n-1)$, we can prove it for n by induction. In fact, defining the constant L and the function $F(x)$ by the following equations :—

$$f(a+h-0)-f(a)-hf'(a)-\ldots-\frac{h^{n-1}}{(n-1)!}f^{(n-1)}(a)=h^nL \ \ldots\ldots(1),$$
$$f(x)-f(a)-(x-a)f'(a)-\ldots$$
$$\ldots-\frac{(x-a)^{n-1}}{(n-1)!}f^{(n-1)}(a)-L(x-a)^n=F(x) \ \ldots(2),$$

we get, as before, a point x_1 such that

$$F'(x_1)=f'(x_1)-f'(a)-\ldots-\frac{(x_1-a)^{n-2}}{(n-2)!}f^{(n-1)}(a)-nL(x_1-a)^{n-1}=0 \ \ldots(3).$$

Now by hypothesis the theorem holds for the integer $(n-1)$ and the function $f'(x)$. Hence there is a point x_2 between x and a such that

$$f^{(n)}(x_2)=n!\,L.$$

Substituting for L in (1) the theorem now follows, if it has been proved true for $n=1$ and $n=2$.

But when $n=2$ we are given that $f'(a)$ exists, while $f'(x)$ is not necessarily continuous at $x=a$, and (3) becomes

$$F'(x_1)=f'(x_1)-f'(a)-2L(x_1-a)=0.$$

Hence, since $f'(a)$ is *one* of the limits of $f'(x)$ as x approaches a, this gives by the above modified Theorem of the Mean,

$$(x_1-a)f''(x_2)-2L(x_1-a)=0,$$

whence
$$f''(x_2)=2L,$$

which proves the theorem for $n=2$, and completes the induction.

27. The Fractional Theorem of the Mean,
$$\frac{f(b-0)-f(a+0)}{F(b-0)-F(a+0)}=\frac{f'(x_1)}{F'(x_1)}.$$

The following extension of the Theorem of the Mean is of considerable importance and includes it as a particular case.

THEOREM. *If $f(x)$ and $F(x)$ be functions which have differential coefficients at every point of the open interval $(a < x < b)$, these differential* *coefficients having in the open interval no common zeros or infinities, then*

$$\frac{f(b-0)-f(a+0)}{F(b-0)-F(a+0)} = \frac{f'(x_1)}{F'(x_1)}$$

for some point x_1 of the open interval. Here the sequences of values of x yielding the limits $f(b-0)$ and $f(a+0)$ are supposed to be the same as those yielding the limits $F(b-0)$ and $F(a+0)$ and the meaning otherwise is that of § 26.

CASE 1.   Let $f(b-0) - f(a+0)$ be not zero.

Put $g(x) \equiv f(x) - f(a+0) - m[F(x) - F(a+0)]$,

where m denotes the left-hand side of the equality to be proved.

Evidently $g(b-0) = g(a+0) = 0$.

Moreover since $f'(x)$ and $F'(x)$ are never infinite together, $g(x)$ is certainly differentiable, as well as continuous, throughout the open interval.

Hence (§ 26), for some value x_1 of the completely open interval,

$$g'(x_1) = 0,$$

that is $0 = f'(x_1) - mF'(x_1).$

This equation would give us no information if f' and F' were known to have common zeros, as x_1 would presumably be one of these zeros. As such common zeros do not exist, it is evident that neither f' nor F' can be zero at $x = x_1$, and, as they have no common infinities, it is equally evident that their values when $x = x_1$ are not infinite. Dividing then the last equation by $F'(x)$, we get the equality to be proved.

CASE 2.   Let $f(b-0) - f(a+0)$ be zero.

Then a point x_1 exists, by § 26, at which $f'(x_1)$ is zero, and $F'(x_1)$, by the hypothesis, is not zero. Thus the theorem is still true.

In the above investigation we have not, in the statement of the theorem, expressly excluded the possibility of $F(b-0) - F(a+0)$ vanishing, because, with the ordinary symbolism, the fraction m would then not have any existence.   Of course, however, if $f(b-0) - f(a+0)$ does not vanish, both sides of the equation to be proved assume the same form $c/0$, where c has a value necessarily different from zero, and in general different on the two sides of the equation.   It is of importance to remark that, if $f(b-0) - f(a+0)$ and $F(b-0) - F(a+0)$ are both zero, then an internal point of the interval can be found such that $f'(x)/F'(x)$ has there any assigned value we please.   In fact the above argument then applies, if m, instead of the meaning there attached to it, is supposed to have the value in question.

28. Other Remainder Forms of Taylor's Theorem; Schlömilch-Roche's, Lagrange's and Cauchy's Forms. We can obtain yet another extension of the Theorem of the Mean, which combines to a certain extent the results of the last two articles, and is not without importance in the theory of Taylor's series. This theorem is as follows:

THEOREM $^{(\pi)}$. *If $f(x)$ and all its differential coefficients up to the $(n-1)th$ inclusive exist and are continuous in the closed interval $(a \leqslant x \leqslant a + h)$, while the n-th differential coefficient exists in the completely open interval $(a < x < a + h)$, and if $F(x)$ is continuous in the whole closed interval and differentiable throughout the open interval in such a manner that $F(x)$ and $f(x)$ have no common zeros or infinities in that open interval, then*

$$f(a + h) = f(a) + hf'(a) + \ldots + \frac{h^{n-1}}{(n-1)!} f^{(n-1)}(a)$$

$$+ \frac{F(a+h) - F(a)}{F'(a+\theta h)} \cdot \frac{h^{n-1}(1-\theta)^{n-1}}{(n-1)!} f^{(n)}(a+\theta h), \qquad (0 < \theta < 1).$$

In all but the last term on the right-hand side write y for a and $b - y$ for h, and denote the sum of these terms by $g(y)$, and denote by R_n the excess of the left-hand side over the right-hand side omitting the last term.

Then we have only to prove that the last term is equal to R_n.

Evidently we have

$$R_n = g(b) - g(a),$$

and therefore

$$\frac{R_n}{F(b) - F(a)} = \frac{g(b) - g(a)}{F(b) - F(a)}.$$

But $\qquad g'(y) = (b-y)^{n-1} f^{(n)}(y)/(n-1)!.$

Hence, applying the theorem of the last article, we get the required result.

COR. *By putting $F(x) = (a + h - x)^r$, we get the well-known Remainder Form of Taylor's Theorem due to Schlömilch and Roche, in which the last term, or remainder, is*

$$\frac{h^n(1-\theta)^{n-r}}{(n-1)! \, r} f^{(n)}(a+\theta h).$$

If in this form we put $r = 1$, we get that due to Cauchy, and if we put $r = n$ that due to Lagrange (§ 26).

29. The Remainder Form of Taylor's Theorem for several variables. We now proceed to give an extended form of the Theorem of the Mean involving differential coefficients of the nth order, applicable to a function of more than one variable.

THEOREM. *If $f(x, y)$ possesses all its differentials up to the $(n-1)$th inclusive in a closed neighbourhood of the point (a, b), and an n-th differential exists in the open neighbourhood, that is, the closed neighbourhood omitting the point (a, b), then, the point (x, y) being in that open neighbourhood,*

$$f(x, y) = f(a, b) + [(x-a) f_a + (y-b) f_b] + \ldots$$
$$+ \frac{1}{(n-1)!} \left(\frac{\partial^{n-1} f}{\partial a^{n-1}}, \ldots \right)\!\!\!\!\Big) x-a, \ y-b \Big)^{n-1}$$
$$+ \frac{1}{n!} \left(\frac{\partial^{n} f}{\partial \xi^{n}}, \ldots \right)\!\!\!\!\Big) x-a, \ y-b \Big)^{n},$$

where (ξ, η) is an internal point of the stretch from (a, b) to (x, y).

To prove this, consider the function $f(a+ut, b+vt) = F(t)$ say, where u and v are constants for the purposes of the investigation. Then we may apply the one-dimensional Taylor's Theorem in Lagrange's Remainder Form (§ 26) to the function $F(t)$, defined as follows :—

$$F(t) = F(0) + tF'(0) + \ldots + \frac{1}{(n-1)!} t^{n-1} F^{n-1}(0) + \frac{1}{n!} t^{n} F^{n}(\tau),$$

where $\qquad\qquad\qquad 0 < \tau < t.$

But, by § 21,

$$F'(t) = u \frac{\partial f}{\partial x} + v \frac{\partial f}{\partial y},$$

$$F''(t) = \left(u \frac{\partial}{\partial x} + v \frac{\partial}{\partial y} \right)^{2} f,$$

$$\ldots \qquad \ldots \qquad \ldots \qquad \ldots$$

$$F^{r}(t) = \left(u \frac{\partial}{\partial x} + v \frac{\partial}{\partial y} \right)^{r} f.$$

To find the values of these expressions when $t = 0$, we have obviously only to write a for x and b for y. Hence

$$F(t) = f + t \left(u \frac{\partial}{\partial a} + v \frac{\partial}{\partial b} \right) f + \ldots + \frac{1}{(n-1)!} t^{n-1} \left(u \frac{\partial}{\partial a} + v \frac{\partial}{\partial b} \right)^{n-1} f$$
$$+ \frac{1}{n!} t^{n} \left(u \frac{\partial}{\partial \xi} + v \frac{\partial}{\partial \eta} \right)^{n} f(\xi, \eta).$$

Hence, finally, putting $ut = x - a$, $vt = y - b$, we get the required result.

It should be noticed that this is a much more general result than that usually given under the name of Cauchy's Theorem, which requires the continuity of all the nth differential coefficients. We have used Lagrange's form of the remainder theorem for one variable; the same argument, word for word, applies if we take the Schlömilch-Roche form (§ 28). It is hardly necessary to add that the whole argument applies whatever be the number of variables (see below § 32).

XI. IMPLICIT FUNCTIONS.

30. Existence and differentials of an implicit function defined by a single equation. So far all the functions that have been considered are supposed to be so defined that there could be no question as to their existence, or as to that of the differential coefficients which occur in the various theorems. Moreover, these differential coefficients are supposed capable of determination by the ordinary simple rules.

We now consider an important class of functions of a single variable defined by the equation to zero of a function of two variables.

THEOREM. *If $f(x, y)$ be a function of x and y, whose value at the point (a, b) is zero, and which in a certain closed neighbourhood of the point (a, b) is continuous with respect to x and with respect to y, and possesses at the same point a finite differential coefficient with respect to y, viz. f_b which is different from zero, then a function y of x exists with the following properties:*

(1) *Its value is b when x is a.*

(2) *When substituted in $f(x, y)$ it makes $f(x, y)$ zero throughout a certain neighbourhood of the point (a, b).*

(3) *Further, if, for each fixed value of x, f is throughout some closed neighbourhood of the point (a, b) a monotone never constant function of y, this function y of x is unique.*

(4) *Further, if f is a continuous function of the ensemble (x, y) in the closed neighbourhood of the point (a, b), this function y of x is a continuous function of x.*

(5) *Finally, if f possesses at the point (a, b) a first differential, then*

this function y of x possesses at the point $x = a$ a first differential coefficient, whose value p is given by

$$f_a + pf_b = 0.$$

For definiteness it will be assumed that f_b is positive. It then follows from the definition of a differential coefficient, and the fact that $f(a, b)$ is zero, that we can find a stretch on the ordinate of the point (a, b), having that point as centre, such that in the upper half f is greater than zero, and in the lower half it is less than zero. Hence also, since f is continuous with respect to x, we can draw two stretches* parallel to the axis of x, having the end-points of the stretch just found for centres, and of equal length, such that in the upper one f is greater than zero, and in the lower one f is less than zero.

Completing this rectangle† we obtain a closed neighbourhood of the point (a, b), such that in it on each ordinate f has a positive value and a negative value, and therefore, since f is continuous with respect to y, f assumes the value zero at one or more points forming a closed set (§ 7).

The y-coordinate of the lowest of these points on each ordinate constitutes a function y of x having the properties (1) and (2).

If f is for each fixed value of x a monotone never constant function of y, this function y of x is unique, for f then only assumes each of its values once on each ordinate, in particular, f is zero once only. This proves (3).

To prove (4), we only have to notice that, if f is continuous with respect to the ensemble (x, y), the plane‡ set of all its zeros in the closed neighbourhood chosen forms a closed set. Hence, taking any sequence $x_1, x_2, \ldots$ having x as limit, the corresponding zeros have as limit the zero on the limiting ordinate, so that the function y of x has for every value of x a value equal to the unique limit of values in the neighbourhood, i.e. it is a continuous function of x.

Finally, if f has a first differential at the point (a, b), there is a closed neighbourhood of the point such that throughout it

$$f(a + h, b + k) = h(f_a + e_1) + k(f_b + e_2),$$

where the e's have zero as limit when h and k approach zero in any manner. Hence, inserting for $b + k$ our function y,

$$0 = (x - a)(f_a + e_1) + (y - b)(f_b + e_2),$$

---

* In the corresponding $(n + 1)$-dimensional discussion these are what may be called hyper-stretches: thus for three dimensions they are squares, for four dimensions cubes, and so on.

† $(n + 1)$-dimensional parallelepiped.　　　　　　　　‡ $(n + 1)$-dimensional.

so that
$$\frac{y-b}{x-a} = -\frac{f_a + e_1}{f_b + e_2},$$
where, when x approaches a in any manner, y being continuous has the limit b, and therefore e_1 and e_2 both have the limit zero.

Proceeding to the limit, we have, by the definition of a differential coefficient,
$$\frac{dy}{dx} = -f_a/f_b,$$
that is,
$$f_a + pf_b = 0.$$

[NOTE. If the condition which secures the uniqueness be omitted, it is evident that the uppermost of all the functions y of x which make $f = 0$ is upper-semi-continuous, and that the lowest is lower-semi-continuous.]

COR. 1. *We may replace the condition that throughout the neighbourhood f should be for each fixed value of x a monotone nowhere constant function of y, by the condition that f_y should exist throughout a closed neighbourhood of (a, b) and be nowhere zero.*

In fact, if f_y is nowhere zero in the neighbourhood it has always the same sign on each ordinate, since for each fixed value of x it assumes on the corresponding ordinate every value between its upper and lower bounds, and therefore could not have opposite signs without being somewhere zero.

It is for the rest clear by applying the Theorem of the Mean that, if f vanishes at two points on an ordinate f_y must vanish at some point between the two points, so that the condition in question necessarily excludes this possibility.

COR. 2. *We may replace the condition in question by the following:— that f_y should exist throughout a closed neighbourhood of the point (a, b), and be continuous at that point with respect to the ensemble (x, y).*

For in this case we can assign a closed neighbourhood of the point (a, b) throughout which f_y has the same sign as at (a, b), and therefore never vanishes, so that we can apply Cor. 1.

31. THEOREM. *If $f(x, y)$ is a function of the ensemble (x, y) which is zero at (a, b), and possesses there an n-th differential, where n is greater than unity, then, provided f_y is not zero at the point (a, b), we can find a closed neighbourhood of the point (a, b), in which there is one, and only one, function $g(x)$ of x which has the value b when $x = a$ and when substituted for y makes $f(x, y)$ identically zero. Further, this function possesses an n-th differential coefficient at the point (a, b) which* *may be obtained by equating to zero the successive total differential coefficients of the function $f(x, y)$, obtained by the ordinary rule.*

First, to prove the theorem when $n = 2$, we remark that, since the function has a second differential at the point (a, b), a first differential exists at and in a closed neighbourhood of the point (a, b). Also since f has a second differential at the point (a, b), f_y has a first differential there, and is therefore continuous at (a, b). Hence, remembering that f_b is different from zero, we may so choose our neighbourhood that f_y is different from zero at every point considered. The neighbourhood so chosen is then such that the conditions of § 30 are satisfied at every point. Hence corresponding to each point x there is a "tile," that is, a closed rectangle with x as centre, and in this tile a unique function $g(x)$, such that $f[x, g(x)]$ is identically zero throughout the tile, and further

$$f_x + g'(x)f_y = 0 \quad \dots\dots\dots\dots\dots\dots\dots(1),$$

or, symbolically, $\left(\dfrac{\partial}{\partial x} + g' \dfrac{\partial}{\partial y} \right) f(x, y) = 0,$

where, after differentiating, we have to insert $g(x)$ for y.

These tiles overlap, but, since in each tile the function $g(x)$ is unique, it follows that the value of $g(x)$ is independent of the particular tile used in determining its value, and is the same whether or no that tile was the one with x as centre. Thus we have an unique function $g(x)$ defined throughout the whole neighbourhood, and it has a differential coefficient at each point, given by the identical equation (1).

Now, since f has a second differential at the point (a, b), f_x and f_y both have first differentials there, and consequently have total differential coefficients with respect to x when we replace y by $g(x)$. Also the right-hand, and therefore the left-hand, side of (1) has a total differential coefficient whose value is zero. Thus, since f_y is not zero at (a, b), we may apply § 21, and say that $g'(x)$ has a first differential coefficient at the point (a, b), and that it is given by totally differentiating the identity (1) with respect to x, and putting $x = a$, $y = b$. That is, $g(x)$ possesses a second differential coefficient, and it is given by

$$f_{aa} + 2f_{ab}g'(a) + f_{bb}\{g'(a)\}^2 + f_b g''(a) = 0,$$

or, say, symbolically

$$\left(\frac{\partial}{\partial x} + g' \frac{\partial}{\partial y} \right)^2 f(x, y) = 0, \quad (x = a, \ y = b),$$

that is, it is obtained by equating to zero the total differential coefficient of the second order of $f(x, y)$ with respect to x, when $y = g(x)$.

It will be noticed that we have, in performing the total differentiation, made no distinction between f_{ab} and f_{ba}, in accordance with the results already proved (§ 19). This proves the theorem when $n = 2$.

Again, if $n = 3$, not only, as we saw, is f_y different from zero at each point of the neighbourhood of (a, b), but at each such point f has a second differential, so that the above reasoning applies, and we may assert that at each such point $g(x)$ has a second differential given by

$$\left(\frac{\partial}{\partial x} + g'\frac{\partial}{\partial y}\right)^2 f(x, y) = 0,$$

that is, by $\qquad (f_{xx}, \ldots \emptyset 1, g')^2 + f_y g''(x) = 0 \qquad \ldots\ldots\ldots\ldots(2).$

The reasoning by which we now deduce the existence and value of $g'''(a)$ from (2) is precisely similar to that by which we deduced the existence and value of $g''(a)$ from (1) in the case when $n = 2$.

For, since f has a third differential at (a, b), f_{xx}, f_{xy} and f_{yy} have first differentials there. Also $g'(x)$ has a first differential coefficient, and therefore a first differential at the same point. Since the product and the sum of functions having a first differential at (a, b) is a function having a first differential there, it follows that the quadratic $(f_{xx}, \ldots \emptyset 1, g')^2$ has a first differential at (a, b). Moreover f_y has a first differential at the same point. Hence both these functions $(f_{xx}, \ldots \emptyset 1, g')^2$ and f_y have total differential coefficients with respect to x, when we replace y by $g(x)$. Again the right-hand side, and therefore the left-hand side, of (2) has the total differential coefficient zero. Thus, since f_b is different from zero, we may apply § 21, and say that $g''(x)$ has a first differential coefficient at the point (a, b), and that it is given by totally differentiating the identity (2) with respect to x, and putting $x = a$, $y = b$. That is, $g'''(a)$ exists, and is given by

$$(f_{aaa}, \ldots \emptyset 1, g')^3 + 3\{f_{ab} + g'(a)f_{bb}\} g'' + f_b g''' = 0,$$

or, symbolically, by

$$\left(\frac{\partial}{\partial x} + g'\frac{\partial}{\partial y}\right)^3 f(x, y) = 0, \quad (x = a, \ y = b).$$

This proves the theorem for $n = 3$. We have now only to notice that, if we have proved the theorem for $n = r$, it follows by corresponding reasoning that it is true for $n = r + 1$. Hence, by induction, the truth of the theorem follows.

32. Equations giving the successive differential coefficients of an implicit function. The equations, giving in order the successive differential coefficients, when written out at length are as follows:

(1) $\left(\dfrac{\partial}{\partial x} + y'\,\dfrac{\partial}{\partial y}\right) f = 0,$ or $f_x + y'f_y = 0$;

(2) $\left(\dfrac{\partial}{\partial x} + y'\,\dfrac{\partial}{\partial y}\right)^2 f = 0,$ or $(f_{xx}, f_{xy}, f_{yy} \,\mathcal{Q}\, 1, y')^2 + f_y y'' = 0$;

(3) $(f_{xxx}, \ldots \mathcal{Q}\, 1, y')^3 + 3\,(f_{xy} + y'f_{yy})\,y'' + f_y y''' = 0$;

(4) $(f_{xxxx}, \ldots \mathcal{Q}\, 1, y')^4 + 6\,(f_{xxy}, \ldots \mathcal{Q}\, 1, y')^2 y'' + 3 f_{yy} y''^2$

$$+ 4\,(f_{xy} + y'f_{yy})\,y''' + f_y y^{(4)} = 0 ;$$

(5) $\left(\dfrac{\partial^5 f}{\partial x^5}, \ldots \mathcal{X}\, 1, y'\right)^5 + 10\,(f_{xxxy}, \ldots \mathcal{Q}\, 1, y')^3 y'' + 10\,(f_{xxy}, \ldots \mathcal{Q}\, 1, y')^2 y'''$

$$+ 5\,(f_{xy}, \ldots \mathcal{Q}\, 1, y')\,y^{(4)} + 15\,(f_{xyy}, \ldots \mathcal{Q}\, 1, y')\,y''^2 + 10 f_{yy} y'' y''' + f_y y^{(5)} = 0,$$

and so on.

It should be noticed that it follows from the mode of formation of these equations that the coefficients are partial differential coefficients of f such that, if any coefficient involves r differentiations with respect to y, it is multiplied by precisely r differential coefficients of y.

33. Implicit function of two or more variables. If now in § 30 we interpret the symbol x to mean the ensemble $(x_1, x_2, \ldots, x_n)$, so that $f(x, y)$ means $f(x_1, x_2, \ldots, x_n, y)$, the theorem becomes a theorem in the theory of functions of $(n + 1)$ variables, no alteration in the wording being required except in the last clause which should now read as follows:

(5) *Finally, if f possesses at the point (a, b) a first differential, then at the point $x = a$ this function y of x possesses a first differential, whose value dy is given by the symbolic equation*

$$\left(h\,\frac{\partial}{\partial x} + dy\,\frac{\partial}{\partial y}\right) f = 0$$

(*that is, written out in full,*

$$h_1 f_{x_1} + h_2 f_{x_2} + \ldots + h_n f_{x_n} + dy f_y = 0),$$

in which the x's are to be replaced by a's and y by b.

The proof of this theorem is almost word for word the same as before; the insignificant verbal alterations have been already given in footnotes.

34. Equations giving the differentials of such an implicit function. Article 31 may similarly be interpreted in space of $(n + 1)$ dimensions. We merely have to change "differential coefficient" in the enunciation into "differential." It is unnecessary to reproduce the proof.

The r-th differential of y at the point a is given symbolically by the equation

$$\left(h \frac{\partial}{\partial x} + dy \frac{\partial}{\partial y} \right)^r f = 0,$$

just as, in the simple case when there is only one x, it was given by

$$\left(\frac{\partial}{\partial x} + y' \frac{\partial}{\partial y} \right)^r f = 0.$$

This equation when expanded has precisely the same form as in § 32, the successive differential coefficients of y being replaced by the successive differentials dy, d^2y, d^3y, ..., $d^r y$, and differentiation with respect to x being replaced by the operator $h \frac{\partial}{\partial x}$ which gives the partial differential with respect to the x's alone.

35. Plurality of solutions. It should be noticed that in § 30, although the uniqueness of the solution is made use of in the proof of the continuity of y, considered as a function of x, in the neighbourhood of the point $x = a$, each solution, even when not unique, is continuous for any value of x such that on the corresponding ordinate there is only one zero of $f(x, y)$, the neighbourhood being chosen sufficiently small; in particular, this is the case at the point $x = a$ itself, if there is no sequence of values of y with b as limit for each of which $f(a, y) = 0$. This follows from the reasoning used in the proof of § 30. If this be the case, the reasoning used in the proof of the property (5) still applies whether or not the solution is unique; that is to say, if $f(x, y)$ has a first differential at (a, b), each of the solutions has a differential coefficient at the point $x = a$, and the value of this differential coefficient is the same for all solutions and is given by

$$f_a + f_b \frac{dy}{dx} = 0.$$

An interesting application of this is constituted by the following theorem.

36. Change of the dependent into the independent variable; $\dfrac{dx}{dy} \cdot \dfrac{dy}{dx} = 1$.

THEOREM. *If $g(x)$ is a function of x which has the value b when $x = a$, and possesses at the point $x = a$ a finite non-vanishing differential coefficient, then there exists at least one inverse function x of y which has the value a when $y = b$, and in a certain closed neighbourhood of the point $y = b$ renders the equation*

$$y = g(x)$$

an identity; further, at the point $y = b$ all these functions possess a common differential coefficient whose value is $1/g'(a)$.

To prove this, we have, in fact, only to put

$$f(x, y) = y - g(x),$$

and apply the above. For, when $x = a$, there is no value of y other than b which makes $f(x, y)$ vanish.

37. Case of uniqueness of the inverse function. Existence of its higher differential coefficients. It follows at once from § 30, or is otherwise evident, that, if the function $g(x)$, mentioned in the enunciation of the preceding theorem, is, in a closed neighbourhood of the point $x = a$, a monotone function of x, the function x of y is unique. In this case $\dfrac{dx}{dy}$ exists at the point, even if $\dfrac{dy}{dx}$ is zero there, being, in fact, infinite with determinate sign.

An important case in which $f(x)$ is monotone is that in which $g'(x)$ exists in the neighbourhood of $x = a$, and is nowhere zero. Moreover applying § 31 we get the following theorem:

THEOREM. *If $g(x)$ is a function of x which has the value b when $x = a$, and possesses at the point $x = a$ a finite n-th differential coefficient where $n \geqslant 2$, and if $g'(a)$ is different from zero, there is one, and only one, function x of y which has the value a when $y = b$, and renders the equation*

$$y = f(x)$$

an identity in a certain closed neighbourhood of the point $y = b$. Moreover this function x of y possesses an n-th differential coefficient at the point.

38. Existence and differentials of implicit functions defined by two or more equations.

THEOREM. *If $f_1(x, y)$, $f_2(x, y)$, $...,f_r(x, y)$ are r functions of the m variables x and the r variables y, which are zero when the x's are equal to a's and the y's to b's, i.e., at the point (a, b), and have n-th* *differentials there, where $n \geqslant 2$, and if the Jacobian J of the f's with respect to the y's*

$$\begin{vmatrix} \dfrac{\partial f_1}{\partial y_1}, & \cdots, & \dfrac{\partial f_r}{\partial y_1} \\[2mm] \dfrac{\partial f_1}{\partial y_2}, & \cdots, & \dfrac{\partial f_r}{\partial y_2} \\[2mm] \cdots\cdots\cdots\cdots \\[1mm] \dfrac{\partial f_1}{\partial y_r}, & \cdots, & \dfrac{\partial f_r}{\partial y_r} \end{vmatrix}$$

is not zero at the point (a, b), then there exist unique functions $y_1, y_2, \ldots, y_r$ of the m variables x, which have the values $b_1, b_2, \ldots, b_r$ at the point a, and, throughout a closed neighbourhood of that point, make all the f's identically zero. Moreover these functions y have n-th differentials at the point a, whose values may be obtained from equations, which in symbolic form are

$$\left(h \frac{\partial}{\partial x} + dy \frac{\partial}{\partial y} \right)^i f_j = 0, \quad (i = 1, 2, \ldots, n \, ; \, j = 1, 2, \ldots, r).$$

Since the Jacobian J is not zero, at least one of all its principal minors is not zero, and we may assume, without loss of generality, that it has been so arranged that the leading principal minor is not zero. It then follows that one of the principal minors of that principal minor is not zero, and we may assume that the determinant is so arranged that that principal minor is the leading minor. Proceeding thus, we may assume that the determinant has been so arranged that none of the leading minors,

$$J_1 = \frac{\partial f_1}{\partial y_1}, \quad J_2 = \begin{vmatrix} \dfrac{\partial f_1}{\partial y_1}, & \dfrac{\partial f_2}{\partial y_1} \\[2mm] \dfrac{\partial f_1}{\partial y_2}, & \dfrac{\partial f_2}{\partial y_2} \end{vmatrix}, \quad J_3 = \begin{vmatrix} \dfrac{\partial f_1}{\partial y_1}, & \dfrac{\partial f_2}{\partial y_1}, & \dfrac{\partial f_3}{\partial y_1} \\[2mm] \dfrac{\partial f_1}{\partial y_2}, & \dfrac{\partial f_2}{\partial y_2}, & \dfrac{\partial f_3}{\partial y_2} \\[2mm] \dfrac{\partial f_1}{\partial y_3}, & \dfrac{\partial f_2}{\partial y_3}, & \dfrac{\partial f_3}{\partial y_3} \end{vmatrix} \cdots,$$

are zero at the point (a, b). Then, since

$$\frac{\partial f_1}{\partial b_1} \neq 0,$$

and f_1 has an nth differential at the point (a, b), where $n \geqslant 2$, there is one, and only one, function y_1 of the remaining $(m + r - 1)$ variables x and y which has the value b_1 at the point $(a_1, \ldots, a_m, b_2, \ldots, b_r)$, and, in a certain closed neighbourhood of that point, makes f_1 identically zero. Moreover, this function y_1 has an nth differential at the point in question.

Now, replace y_1 in f_2 by the function so found, and call the result F_2. This is a function of the m variables x and the remaining $(r-1)$ variables y, which has the value zero at the point $(a_1, \ldots, a_m, b_2, \ldots, b_r)$, since $y_1 = b_1$ there.

Also it has an nth differential at the same point since f_1 and y_1 both have nth differentials. Also $\dfrac{\partial F_2}{\partial y_2}$ is not zero at the point, since, F_2 having a differential with respect to the x's and y's, and therefore with respect to the y's, at the point and in its neighbourhood,

$$\frac{\partial F_2}{\partial y_2} = \frac{\partial f_2}{\partial y_2} + \frac{\partial f_2}{\partial y_1}\frac{\partial y_1}{\partial y_2},$$

where

$$\frac{\partial f_1}{\partial y_2} + \frac{\partial f_1}{\partial y_1}\frac{\partial y_1}{\partial y_2} = 0,$$

so that

$$\begin{vmatrix} -\dfrac{\partial F_2}{\partial y_2} + \dfrac{\partial f_2}{\partial y_2}, & \dfrac{\partial f_2}{\partial y_1} \\[2ex] \dfrac{\partial f_1}{\partial y_2}, & \dfrac{\partial f_1}{\partial y_1} \end{vmatrix} = 0,$$

or

$$J_1 \frac{\partial F_2}{\partial y_2} + J_2 = 0.$$

Since J_1 and J_2 are both different from zero at the point, this gives a finite value different from zero for $\dfrac{\partial F_2}{\partial b_2}$.

We can therefore again apply § 31, and assert that one, and only one, function y_2 of the m variables x and the remaining $(r-2)$ variables y exists, which has the value b_2 at the point $(a_1, \ldots, a_m, b_3, \ldots, b_r)$, and in a certain closed neighbourhood of that point makes F_2 identically zero. Moreover this function y_2 has an nth differential at the point.

Now, the effect of inserting this function y_2 in F_2 is obviously the same as that of inserting this y_2 in the y_1 previously found, and then in f_2 inserting this last y_1, and the function y_2 wherever it occurs explicitly. Doing this both in f_1 and in f_2 they become functions of the $(m+r-2)$ variables which vanish identically. Making the same change in f_3 and denoting the result by F_3, we thus get the three equations

$$\frac{\partial F_3}{\partial y_3} = \frac{\partial f_3}{\partial y_3} + \frac{\partial f_3}{\partial y_2}\frac{\partial y_2}{\partial y_3} + \frac{\partial f_3}{\partial y_1}\frac{\partial y_1}{\partial y_3},$$

$$0 = \frac{\partial f_2}{\partial y_3} + \frac{\partial f_2}{\partial y_2}\frac{\partial y_2}{\partial y_3} + \frac{\partial f_2}{\partial y_1}\frac{\partial y_1}{\partial y_3},$$

$$0 = \frac{\partial f_1}{\partial y_3} + \frac{\partial f_1}{\partial y_2}\frac{\partial y_2}{\partial y_3} + \frac{\partial f_1}{\partial y_1}\frac{\partial y_1}{\partial y_3},$$

holding at the point in question, whence

$$J_2 \frac{\partial F_3}{\partial y_3} + J_3 = 0.$$

Thus $\frac{\partial F_3}{\partial y_3}$ is finite and different from zero at the point

$$(a_1, \ldots, a_m, b_4, \ldots, b_r),$$

at which, as before, F_3 vanishes, and has an nth differential. Thus we can again apply the theorem and deduce the existence of one, and only one, function y_3 of the $(m + r - 3)$ variables having the value b_3 at the point, and in a certain closed neighbourhood of it making F_3 identically zero. This process may be continued and it is evident that we shall at each stage obtain unique functions $y_1, y_2, \ldots, y_i$ of the remaining $(m + r - i)$ variables having the proper values at the point considered, and having nth differentials there, and making $f_1, f_2, \ldots, f_i$ vanish identically in a certain neighbourhood of the point. Inserting these values in f_{i+1} and denoting the result by F_{i+1}, we then clearly get the equations

$$\frac{\partial F_{i+1}}{\partial y_{i+1}} = \frac{\partial f_{i+1}}{\partial y_{i+1}} + \frac{\partial f_{i+1}}{\partial y_i}\frac{\partial y_i}{\partial y_{i+1}} + \ldots + \frac{\partial f_{i+1}}{\partial y_1}\frac{\partial y_1}{\partial y_{i+1}},$$

$$0 = \frac{\partial f_{i-r}}{\partial y_{i+1}} + \frac{\partial f_{i-r}}{\partial y_i}\frac{\partial y_i}{\partial y_{i+1}} + \ldots + \frac{\partial f_{i-r}}{\partial y_1}\frac{\partial y_1}{\partial y_{i+1}}, \quad (r = 0, 1, \ldots i - 1),$$

holding at the point, so that

$$J_i \frac{\partial F_{i+1}}{\partial y_{i+1}} + J_{i+1} = 0.$$

Thus again we can apply the theorem and proceed a stage further. This may be continued until we have exhausted all the y-coordinates, when we shall have expressed each of them in one, and only one, way, so as to have the values b at the point a, and, in a certain closed neighbourhood of that point, to make the r functions f vanish identically; moreover, these functions y of the x's have nth differentials at the point. This being so we have only to form the total differentials of the functions f with respect to the x's, regarding the y's as being these functions, and equate the result to zero, to obtain equations which determine the values of the differentials of the y's at the point a; these equations may be written, symbolically, in the form

$$\left(h\frac{\partial}{\partial x} + dy\frac{\partial}{\partial y}\right)^i f_j = 0, \quad (i = 1, 2, \ldots, n; j = 1, 2, \ldots, r).$$

THEOREM. *If $f_1(x, y)$, $f_2(x, y)$, $\ldots$, $f_r(x, y)$ are r functions of the m variables x and the r variables y which are zero at the point (a, b),* *and have first differentials at the point* (a, b) *and in a closed neighbourhood of that point first differentials with respect to the y's, and if in that closed neighbourhood the Jacobian J of the f's with respect to the y's is not zero, and the same is true of one of its principal minors* J_{r-1}, *and of one of the principal minors* J_{r-2} *of* J_{r-1}, *and so on, down to one of the common constituents of J,* J_{r-1}, ..., J_2, *then there exist unique functions* y_1, y_2, ..., y_r *of the m variables x, which have the values* b_1, b_2, ..., b_r *at the point a, and, throughout a closed neighbourhood of that point, make all the f's identically zero.   Moreover, these functions y have first differentials at the point a, whose values may be obtained by solving the equations*

$$\left(h\, \frac{\partial}{\partial x} + dy\, \frac{\partial}{\partial y} \right) f_i = 0,$$

that is,

$$h_1 \frac{\partial f_i}{\partial x_1} + h_2 \frac{\partial f_i}{\partial x_2} + \ldots + h_m \frac{\partial f_i}{\partial x_m} + dy_1 \frac{\partial f_i}{\partial y_1} + \ldots + dy_r \frac{\partial f_i}{\partial y_r} = 0,$$

for all integers i from 1 *to r both inclusive.*

The proof of this theorem is essentially the same as that of the preceding theorem, quoting Cor. 1, § 30 instead of the theorem of § 31.

COR. 1.   *The conditions that J,* J_{r-1}, ..., J_2, J_1 *should not vanish in a closed neighbourhood of the point* (a, b), *being replaced by the conditions that they should not vanish at the point and be continuous there, the theorem still holds.*

COR. 2.   *The same conditions being replaced by the conditions that J should not vanish at the point, and all the partial differential coefficients of the f's with respect to the y's should be continuous at the point, the theorem still holds.*

COR. 3.   *The conditions that the f's should have first differentials with respect to the y's in a closed neighbourhood of the point, and that J,* J_{r-1}, ..., J_1 *should not vanish throughout that neighbourhood, may be replaced by the conditions that all the partial differential coefficients of the f's with respect to the y's should be continuous throughout a closed neighbourhood of the point, and the Jacobian J should not be zero at the point, the theorem then still holds.*

A particular case of this is a theorem due to Dini that, if J does not vanish at the point (a, b), and all the partial differential coefficients of the f's are continuous, there is a unique set of solutions y_1, ..., y_r, each of which will then have a first differential at the point.

39. Plurality of solutions. Change of variables. If
in the preceding theorem we only assume that at the point (a, b),
but not necessarily in the neighbourhood, J not zero, it is evident
that we still get at least one set of solutions, the uniqueness of
this set being the only thing affected. An argument similar to that
given in § 35 shews that each such solution is continuous for any
ensemble x, such that there is only one ensemble y for which the given
functions $f_1, f_2, \ldots, f_r$ all vanish. In particular this will be the case at
the point $x = a$, provided the point $y = b$ is not a limiting point of
points y for which $f_1(a, y), f_2(a, y), \ldots, f_r(a, y)$ all vanish.

Supposing this to be the case, the reasoning by which the existence
of the first differentials was demonstrated still holds, and the equations
determining them are the same for all possible sets of solutions.

We thus easily get the following theorem, which corresponds to the
theorem of § 36.

THEOREM. *If $g_1(x), g_2(x), \ldots, g_r(x)$ are functions of the r variables
x which have the values $b_1, b_2, \ldots, b_r$ when the x's have the values
$a_1, \ldots, a_r$, and possess at the point $x = a$ a finite non-vanishing Jacobian
J, then there exists at least one set of functions x of the r variables y
which have the values $a_1, \ldots, a_r$ at the point $y = b$, and which in a certain
closed neighbourhood of that point render the equations*

$$y_1 = f_1(x), \quad y_2 = f_2(x), \quad \ldots, \quad y_r = f_r(x),$$

*identities; further, all these sets of functions possess at the point $y = b$
a common Jacobian J' whose value is $1/J$.*

Moreover, corresponding to the theorem of § 37, we may assert
the uniqueness of the set of solutions, provided the f's have nth
differentials, where $n \geqslant 2$, and it will follow that the solutions themselves
possess nth differentials at the point $x = a$.

40. The relation $\dfrac{dy}{dx} = \dfrac{dy}{dt} \Big/ \dfrac{dx}{dt}$. A particular case of the last
theorem is the following :

*If x and y are both functions of t possessing differential coefficients
at the point $t = t_0$, and $\dfrac{dx}{dt_0} \neq 0$, then there exists at least one function y of
x, and all these functions have at the corresponding point x_0 a common
differential coefficient, viz.*

$$\frac{dy}{dx} = \frac{\dfrac{dy}{dt_0}}{\dfrac{dx}{dt_0}}.$$

If x is a monotone function of t, or if $\dfrac{dx}{dt}$ exists also in the neighbour-hood of the point $t = t_0$ and is nowhere zero, this solution is unique.

XII. ON THE REVERSIBILITY OF THE ORDER OF PARTIAL DIFFERENTIATION.

41. Sufficient conditions for the equivalence of f_{xy} and f_{yx}.

What is in some respects the most important of such sets of conditions has already been given in the Fundamental Theorem of Differentials, viz. the equality holds if $\dfrac{\partial f}{\partial x}$ and $\dfrac{\partial f}{\partial y}$ have at the point in question first differentials. Bearing in mind the corollary of § 18, this at once gives us the following theorem.

THEOREM. *If $\dfrac{\partial^2 f}{\partial x^2}$ and $\dfrac{\partial^2 f}{\partial y^2}$ are both continuous functions of the en-semble (x, y) at the point (a, b), and $\dfrac{\partial^2 f}{\partial x \partial y}$ and $\dfrac{\partial^2 f}{\partial y \partial x}$ both exist at this point, then these latter have equal values there.*

It will be noticed that these conditions make no assumption as to the existence at other points, much less as to the continuity [p], of the mixed partial differential coefficients f_{xy} and f_{yx}.

A third set of sufficient conditions is the following [23]. For convenience we give the statement in two parts.

PART I (On the existence of f_{yx} at the point (a, b)). *If f_x exists in a closed neighbourhood of a point (a, b), while in the completely open neigh-bourhood, excluding the axial cross (§ 5), it has a differential coefficient f_{yx} with respect to y, then, if f_{yx} has only one double limit as we approach the point (a, b) in any manner by means of points not on the axial cross, f_{yx} exists also at the point (a, b) itself.*

For, by a repeated application of the Theorem of the Mean,

$$\frac{f(a+h,\, b+k) - f(a+h,\, b) - f(a,\, b+k) + f(a,\, b)}{hk}$$
$$\equiv m(a,\, b;\ a+h,\, b+k) = f_{yx}(x',\, y'),$$

where the point (x', y') does not lie on the axial cross, and has (a, b) as limiting point when h and k each approach zero in any manner without assuming the value zero.

Since $f_{yx}(x, y)$, and therefore $f_{yx}(x', y')$, has only one double limit at the point (a, b), the same is true of $m(a,\, b;\ a+h,\, b+k)$ when h

and k have zero as limit. Hence, a repeated limit being a double limit (§ 5), all the repeated limits of $m(a, b; a+h, b+k)$ are equal.

But, by hypothesis, f_x is defined on the ordinate $x = a$, so that

$$\frac{f_x(a, b+k) - f_x(a, b)}{k} = \underset{h=0}{\text{Lt}}\ m(a, b; a+h, b+k),$$

and has, therefore, by what has been shewn, only one limit when k has zero as limit; that is to say, f_x has a differential coefficient f_{yx} with respect to y at the point (a, b).

Note. It should be noticed that it does not follow that f_{yx} is continuous at (a, b) with respect to either variable, still less with respect to the ensemble (x, y). In fact, f_{yx} need not exist on the axial cross, except at the point (a, b) itself. It may be proved, however, that when it does exist on the axial cross, it is continuous at (a, b).

PART II. *If in addition to the preceding requirements f_y exists along the line $y = b$ at and in the neighbourhood of the point (a, b), then f_{xy} also exists at the point (a, b) and has the same value as f_{yx}.*

For, in this case,

$$\frac{f_y(a+h, b) - f_y(a, b)}{h} = \underset{k=0}{\text{Lt}}\ m(a, b; a+h, b+k),$$

and has, therefore, as was shewn in the preceding proof, only one limit when h has zero as limit; that is to say, f_y has a differential coefficient f_{xy} with respect to x at the point (a, b).

Since the value of $f_{xy}(a, b)$, like that of $f_{yx}(a, b)$, is thus the unique double limit of $m(a, b; a+h, b+k)$,

$$f_{xy}(a, b) = f_{yx}(a, b). \qquad \text{Q.E.D.}$$

Note 1. It has nowhere been assumed that the unique limit postulated is finite; it may be $+\infty$ or $-\infty$.

Note 2. It should be noticed that, without making any properly two-dimensional hypothesis, *we can prove that f_{yx} exists at the point (a, b) if we postulate that f_{yx} exists and is finite along the ordinate $x = a$ in some open neighbourhood of the point (a, b) but not at the point itself, and that it has an unique limit as we approach the point (a, b) along that ordinate.*

This is an immediate consequence of the Theorem of the Mean for a single variable applied to f_x regarded as a function of y at points of the ordinate $x = a$.

Note 3. In the proof of the existence of an unique double limit for $m(a, b; a+h, b+k)$, the assumption (1) that f_x exists in the open neighbourhood of the point (a, b) excluding the ordinate $x = a$, was rendered necessary in order to apply the Theorem of the Mean.

This being postulated, the further assumption (2) that f_x also exists on the ordinate $x = a$, at and in the neighbourhood of the point (a, b), is needed in order to ascribe a meaning to the expression

$$f_x(a, b + k) - f_x(a, b),$$

and so to prove the existence of f_{yx} at the point (a, b).

Similarly, without postulating (2), the existence of f_{xy} requires the assumption (3) that f_y exists on the line $y = b$, at and in the neighbourhood of the point (a, b).

XIII. POWER SERIES.

42. Continuity and differentiability of power series inside the region of convergence. We now propose to investigate the continuity and differentiability of a function of a single variable defined by an infinite series of positive integral powers. This forms an indispensable preliminary to the discussion of Taylor's Theorem.

Let $\qquad\qquad f(x) = a_0 + a_1 x + a_2 x^2 + \dots \quad \dots\dots\dots\dots\dots(1),$

where the series is convergent for all values of x in the half-open interval $(0 \leqslant x < r)$.

I. *The series is absolutely convergent in this half-open interval.*

For let x be any value in this interval, then we have to prove that the series

$$A_0 + A_1 x + A_2 x^2 + \dots \quad \dots\dots\dots\dots\dots\dots(2)$$

converges, where the A's denote the absolute values of the a's.

Let X be any quantity greater than x and less than r. Then, since (1) converges when $x = X$, the nth term has the limit zero as n increases indefinitely, the same is therefore true of its absolute value. Thus we may write

$$A_n X^n < B,$$

where B is some fixed finite quantity. Therefore

$$A_n x^n < B \left\{ \frac{x}{X} \right\}^n$$

Bearing in mind that x is less than X, this shews that the series (2) has its terms less than the corresponding terms of a convergent geometrical series, which proves the theorem.

II. $f(x)$ *is in the whole open interval a continuous function of x.*

To shew this we shall prove that $f(x)$ is continuous in every closed interval inside this open interval.

Let X be any fixed positive quantity less than r, and x a variable point of the closed interval $(0, X)$. Put

$$f_0(x) = A_0 + A_1X + A_2X^2 + \dots, \qquad \dots\dots\dots(2')$$
$$f_1(x) = a_0 + A_1X + A_2X^2 + \dots,$$
$$f_2(x) = a_0 + a_1x + A_2X^2 + \dots,$$

and so on, $f_n(x)$ agreeing with the series (1) for the first n terms and with the series $(2')$ in all subsequent terms. Then it is clear that these functions form a monotone decreasing sequence of continuous functions and that their limit is $f(x)$.

Now put

$$g_0(x) = -A_0 - A_1X - A_2X^2 - \dots,$$
$$g_1(x) = a_0 - A_1X - A_2X^2 - \dots,$$
$$g_2(x) = a_0 + a_1x - A_2X^2 - \dots,$$

and so on, where $g_n(x)$ agrees with the series (1) for the first n terms and with the series for $g_0(x)$ in all subsequent terms. Then it is clear that these functions g_n form a monotone increasing sequence of continuous functions and that their limit is $f(x)$.

Thus $f(x)$ being the limit both of a monotone increasing and of a monotone decreasing sequence of continuous functions at a point x, it follows that $f(x)$ is continuous at the point x (§ 7).

III. *The series* $\qquad a_1 + 2a_2x + 3a_3x^2 + \dots, \qquad \dots\dots\dots(3)$

and the series $\qquad a_0x + \tfrac{1}{2}a_1x^2 + \tfrac{1}{3}a_2x^3 + \dots, \qquad \dots\dots\dots(4)$

have the same properties in the open interval as the series (1).

In fact, since $n^{1/n}$ has, as n increases, the limit unity*, and x is less than r, we can always find a value of n such that, for this and all greater values,

$$\frac{X}{x} > n^{1/n},$$

where X is any fixed value greater than the fixed value x and less than r. For this and all greater values of n we have, therefore,

$$nx^n < X^n,$$

---

* For, if $10^k < n \leqslant 10^{k+1}$, we have $10^{k/n} < n^{1/n} \leqslant 10^{(k+1)/n}$. Hence, taking logarithms, $k/n < \log n^{1/n} \leqslant (k+1)/n$, so that, by the first inequality,

$$\frac{k}{10^{k+1}} < \log(n^{1/n}) < \frac{k+1}{10^k},$$

which shews that $\log(n^{1/n})$ has the limit zero.

and, therefore,

$$nA_n x^{n-1} < A_n X^n \frac{1}{x};$$

hence, as series $(2')$ is convergent, the series (3) is absolutely convergent, and has therefore the property II.

That the series (4) has the properties in question follows immediately from the fact that each term is less in absolute value than the corresponding term in (1) for the same value of x.

IV. $f(x)$ *has a differential coefficient, and its differential coefficient is the sum of the series* (3) *at every point of the open interval. Further, the function represented by the sum of the series* (4) *has* $f(x)$ *for its differential coefficient at every point of the open interval.*

Bearing in mind that

$$(x+h)^n - x^n$$

is divisible by h, and has for quotient

$$\binom{n}{1} x^{n-1} + \binom{n}{2} x^{n-2} h + \ldots + h^{n-1} = u_n, \text{ say,}$$

we have, by the known rule for the subtraction of convergent series*,

$$\frac{f(x+h) - f(x)}{h} = a_1 u_1 + a_2 u_2 + \ldots + a_n u_n + \ldots \ldots \ldots (5),$$

provided h has been chosen so small that $x + h$, as well as x, lies in the given open interval.

We can therefore suppose X so chosen that

$$x + h < X < r.$$

Thus, since when s is greater than unity,

$$\binom{n}{s} < n \binom{n-1}{s-1},$$

so that

$$u_n < n \left\{ x^{n-1} + \binom{n-1}{1} x^{n-2} h + \ldots + h^{n-1} \right\},$$

we have

$$u_n < n(x+h)^{n-1} < n X^{n-1},$$

and therefore

$$A_n u_n < n A_n X^{n-1}.$$

Hence since by III, the series

$$A_1 + 2A_2 X + 3A_3 X^2 + \ldots \ldots \ldots \ldots \ldots \ldots (6)$$

is convergent, the series (5) is absolutely convergent.

* This rule is an immediate consequence of the fact that the difference of the limits (when they are unique and finite) of two functions is the unique limit of the difference of the two functions.

Carrying out precisely the same process as in II, using for f_0 the series (6), we shew that the series (5) is the limit both of a monotone increasing and of a monotone decreasing sequence of continuous functions of (x, h) and is therefore (§ 7) a continuous function of the ensemble (x, h) at all points for which

$$0 \leqslant x \leqslant x + h < X < r,$$

and, therefore, for each fixed value of x, and for the value zero of h. Hence

$$\frac{f(x+h)-f(x)}{h}$$

has, for each fixed value of x in the open interval considered, an unique limit as h approaches zero, and this limit is got by putting $h = 0$ in the right-hand side of (5). By similar reasoning

$$\frac{f(x-h)-f(x)}{-h}$$

has, as h approaches zero, an unique limit, whose value is again got by putting $h = 0$ in the right-hand side of (5).

This proves the first of our statements. The second statement follows from the fact that the series (4) has, by III, all the properties of the series (1), and therefore, by what has just been proved, has the series (1) for differential coefficient.

43. Continuity on the boundary. All the theorems proved so far refer to the half-open interval $(0 \leqslant x < r)$, where we may suppose r to be the upper bound of the values of x, for which the series (1) converges. When $x = r$ the series may, or may not, converge; if it converges, it may, or may not, converge absolutely. If it does not converge, it may diverge or oscillate. Thus, for instance,

$$x - \tfrac{1}{2}x^2 + \tfrac{1}{3}x^3 - \ldots \text{ converges, but not absolutely,}$$

$$\frac{x^2}{1 \cdot 2} - \frac{x^3}{2 \cdot 3} + \frac{x^4}{3 \cdot 4} - \ldots \text{ converges absolutely,}$$

$$x + \tfrac{1}{2}x^2 + \tfrac{1}{3}x^3 + \ldots \text{ diverges to } + \infty,$$

$$1 - x + x^2 - \ldots \text{ oscillates,}$$

for the value $x = r = 1$.

Notice also that, if we write x^2 for x in the second of these examples, we get a series which converges absolutely in the whole closed interval $(-1, +1)$, and does not converge elsewhere.

As regards the continuity on the boundary, we have the following theorem :

The limits of $f(x)$, as x approaches the value r, lie between the upper and lower limits* of the series

$$a_0 + a_1 r + a_2 r^2 + \dots \qquad \dots\dots\dots\dots\dots(7).$$

In particular, $f(x)$ has, if the series (7) converges or diverges to a definite limit, this limit as unique limit at $x = r$.

Writing $S_0, S_1, S_2, \dots, S_m, \dots,$

for the sum of the first one, two, three, $\dots, (m+1), \dots$ terms of the series (7) we get, after multiplying the absolutely convergent series (1) by the absolutely convergent series

$$1 + \frac{x}{r} + \frac{x^2}{r^2} + \dots,$$

the product series $S_0 + S_1 \frac{x}{r} + S_2 \frac{x^2}{r^2} + \dots,$

whence $f(x) = \left(1 - \frac{x}{r}\right)\left(S_0 + S_1 \frac{x}{r} + S_2 \frac{x^2}{r^2} + \dots\right) \quad \dots\dots\dots(8).$

Hence denoting by l_m and u_m the lower and upper bounds respectively of the first m of the quantities S_i, and by L_m and U_m the lower and upper bounds respectively of the remaining quantities S_i, we have at once

$$f(x) \geqslant \left(1 - \frac{x}{r}\right)\left\{\left(1 + \frac{x}{r} + \dots + \frac{x^{m-1}}{r^{m-1}}\right) l_m + \frac{x^m}{r^m}\left(1 + \frac{x}{r} + \dots\right) L_m\right\}$$

$$\geqslant l_m \left(1 - \frac{x^m}{r^m}\right) + L_m \frac{x^m}{r^m},$$

for all values of x in the open interval. But the right-hand side has the unique limit L_m as x approaches the value r. Hence all the limits of $f(x)$ are greater than or equal to L_m.

This being true for all values of m, it is true for the upper bound of L_m for all values of m, that is for the lower limit L (§ 3) of the series (7).

Similarly all the limits of $f(x)$ are less than or equal to the upper limit U of the same series. In other words,

$$L \leqslant \operatorname*{Llt}_{x=r} f(x) \leqslant U.$$

Here we have tacitly assumed that L_m and U_m, but not necessarily L and U, are finite. Should either be not finite, the corresponding part of the inequality is obvious, since L_m could evidently only be $-\infty$ and U_m only $+\infty$.

---

* That is the upper and lower limits of $(a_0 + a_1 r + \dots + a_n r^n)$ as n is indefinitely increased.

44. Differentiability on the boundary. Assuming that the given series

$$a_0 + a_1 x + a_2 x^2 + \dots \qquad \dots \dots \dots (1)$$

converges for the value $x = r$, we have to answer the questions:

(i) *When does the series*

$$a_0 x + \tfrac{1}{2} a_1 x^2 + \dots \qquad \dots \dots \dots (4)$$

have the given series for left-hand differential coefficient at $x = r$?

The answer is, Always: for the ratio of the nth term of the series (4) to that of the original series has zero as limit, so that the series (4) certainly converges and represents a continuous function up to and including $x = r$.

Since the series (1) is the differential coefficient of (4) for $x < r$, and has at $x = r$ an unique limit, viz. its sum there (§ 43), the series (4) has, by a known property of a differential coefficient (p. 19, § 15, Cor. 3), a left-hand differential coefficient at $x = r$, whose value is that unique limit.

If the original series diverges definitely to $+\infty$ or to $-\infty$ when $x = r$, the same argument applies, *if* the series (4) converges.

(ii) *When does the differential series*

$$a_1 + 2a_2 x + 3a_3 x^2 + \dots$$

represent the left-hand differential coefficient of the original series at $x = r$? The answer already supplied to (i) tells us that this is the case whenever the differential series converges, or diverges definitely to $+\infty$ or to $-\infty$.

XIV. TAYLOR'S THEOREM.

45. Taylor's Theorem. We now pass to the necessary and sufficient condition that a function of a real variable x should be capable of expansion in some fixed neighbourhood of a point a in a series of positive integral powers of $(x - a)$.

We require the following theorem :—*If, for each fixed value of $H < b$,*

$$G_p(x, y ; n) = \left| \frac{y^{n+p}}{(n+p)!} f^{(n)}(a + x) \right|,$$

*where p is any fixed positive or negative integer, or zero, has for all
integers n, for which n + p ⩾ 0 and for all values of x and y satisfying*

$$0 \leqslant x \leqslant x + y \leqslant H < b,$$

a finite upper bound, then the same property is possessed by G_{p-q}, *where
q is any positive or negative integer.*

(N.B. This does not require G_p to have a finite upper bound for all
values of x and y such that

$$0 \leqslant x \leqslant x + y < b.)$$

For, identically, z denoting a positive quantity, presently to be
chosen,

$$G_{p-1}(x, y ; n) = \frac{n + p}{(1 + z/y)^{n+p-1}} \cdot \frac{1}{y + z} G_p(x, y + z ; n)$$

$$\leqslant \frac{m + p}{(1 + z/b)^{m+p-1}} \cdot \frac{1}{z} \cdot G_p(x, y + z ; n) \ldots (1),$$

where m is the first integer such that

$$m + p \geqslant b/z.$$

Choose $z < b - H$, then if in $G_{p-1}(x, y ; n)$ we have all the values
of x and y such that

$$0 \leqslant x \leqslant x + y \leqslant H < b,$$

in $G_p(x, y + z ; n)$ we have values of x and y + z such that

$$0 \leqslant x \leqslant x + y + z \leqslant H' < b, \text{ where } H' = H + z,$$

so that, by the hypothesis, $G_p(x, y + z ; n)$, and therefore the right-hand
side of (1), has a finite upper bound. The same is therefore true of the
left-hand side of (1), that is, G_{p-1} has the property specified in the
enunciation.

Hence G_{p-2}, G_{p-3}, ..., G_{p-q} all have the property, q being any
positive integer.

Again since, identically,

$$G_{p+q} = \frac{y^q}{(n + p + q)(n + p + q - 1) \ldots (n + p + 1)} G_p < b^q G_p,$$

the property is certainly possessed by G_{p+q} for it is known to be
possessed by G_p. This proves the theorem.

46. We can now prove the following theorem, long known under
various forms as Taylor's Theorem :

THEOREM. *If f(a + h) and all its differential coefficients exist
and are finite throughout the half-open interval*

$$0 \leqslant h < b,$$

the necessary and sufficient condition that Taylor's series
$$f(a) + hf'(a) + \tfrac{1}{2}h^2 f''(a) + \dots$$
should converge and have $f(a + h)$ for its sum for all values of h for which
$$0 \leqslant h < b$$
is that
$$\left| \frac{y^{n+p}}{(n+p)!} f^n(a+x) \right|,$$
regarded as a function of the ensemble $(x, y\,;\,n)$, should for any conveniently chosen fixed value of p (either zero or a positive or a negative integer) have a finite upper bound for all values of n for which $n + p \geqslant 0$, and all values of (x, y) satisfying the inequalities
$$0 \leqslant x \leqslant x + y \leqslant H,$$
whatever value of H be chosen less than b.

First to shew that this is sufficient. By the preceding article it follows that, whatever value p may have, we may assume that the hypothesis is also satisfied for a negative integer $-r$, where $r \geqslant 2$. That is,
$$\left| \frac{y^{n-r}}{(n-r)!} f^{(n)}(a+x) \right| \quad \dots\dots\dots\dots\dots\dots(1)$$
is bounded for all values of (x, y) such that
$$0 \leqslant x \leqslant x + y \leqslant H < b.$$
Now put $\quad x = \theta h, \; y = (1 - \theta)h, \;$ where $\; 0 < \theta < 1$.
Then (1) becomes
$$\frac{r(n-1)(n-2)\dots(n-r+1)}{h^r} \; \frac{(1-\theta)^{n-r}h^n f^{(n)}(a+\theta h)}{r \cdot (n-1)!},$$
where the second fraction is Schlömilch's remainder R_n, and
$$f(a+h) = f(a) + hf'(a) + \tfrac{1}{2}h^2 f''(a) + \dots$$
$$+ \frac{1}{(n-1)!} h^{n-1} f^{(n-1)}(a) + R_n \dots\dots(2),$$
since the hypothesis that f and all its differential coefficients exist and are finite at every point of the closed interval $(a, a + H)$ renders the remainder form of Taylor's Theorem valid for every value of n.

By the above, Schlömilch's remainder R_n is numerically less than
$$\frac{b^r U}{n-1},$$
where U is the finite upper bound of (1) for all positive values of $h \leqslant H$. Thus R_n has the unique limit zero when n is indefinitely

increased, no matter what function θ may be of n. Hence, by (2), the infinite series

$$f(a) + hf'(a) + \tfrac{1}{2}h^2 f''(a) + \ldots$$

converges and has $f(a + h)$ for its sum for all positive values of $h \leqslant H$. Since this is true for all values of H less than b, this proves the sufficiency of the condition.

To prove that the condition is necessary, we assume that for all values of h for which

$$0 < h \leqslant H < b,$$

$$f(a + h) = f(a) + hf'(a) + \tfrac{1}{2}h^2 f''(a) + \ldots.$$

Since this power series is absolutely convergent, the same is true of

$$F(h) = |f(a)| + h|f'(a)| + \tfrac{1}{2}h^2 |f''(a)| + \ldots = \sum_{n=0}^{\infty} \frac{1}{n!} h^n |f^{(n)}(a)|,$$

and we may differentiate term by term. We thus get

$$F^{(r)}(h) = \sum_{n=r}^{\infty} \frac{1}{(n-r)!} h^{n-r} |f^{(n)}(a)| \geqslant \left| \sum_{n=r}^{\infty} \frac{1}{(n-r)!} h^{n-r} f^{(n)}(a) \right| \geqslant |f^{(r)}(a+h)|$$

and also

$$F^{(r)}(h + k) = \sum_{n=r}^{\infty} \frac{1}{(n-r)!} (h+k)^{n-r} |f^{(n)}(a)|$$

for all values of h and k such that $0 \leqslant h \leqslant h + k \leqslant H < b$. Rearranging the right-hand side in powers of k, which we may do since all the series concerned are absolutely convergent,

$$F^{(r)}(h + k) = \sum_{n=r}^{\infty} \frac{k^{n-r}}{(n-r)!} F^{(n)}(h) \qquad \ldots\ldots\ldots\ldots\ldots(3).$$

Since each term in (3) is positive or zero, it follows that each term has an upper bound which is not greater than the upper bound of $F^{(r)}(h + k)$. But since $F^{(r)}$ is continuous it assumes its upper bound as one of its values, so that that upper bound is finite, and therefore so is the upper bound of the term

$$\frac{k^{n-r}}{(n-r)!} F^{(n)}(h),$$

still more that of

$$\left| \frac{k^{n-r}}{(n-r)!} f^{(n)}(a + h) \right|,$$

and this is true for all values of $r \geqslant 0$. This proves the necessity of the condition for every positive or zero value of r, and therefore, by the theorem of § 45, for every value of r. Since, finally, H is *any* positive value less than b, the necessity of the condition has been fully established.

47. We have now merely to remark that the upper bound of

$$\left| \frac{k^{n-p}}{(n-p)!} f^{(n)}(a+h) \right|,$$

where h and k are connected by the inequalities

$$0 \leqslant h \leqslant h + k \leqslant H < b$$

regarded as a function of the ensemble $(h, k; n)$, is equal to the upper bound of

$$\left| \frac{(H-h)^{n-p}}{(n-p)!} f^{(n)}(a+h) \right|$$

regarded as a function of the ensemble (h, n). This follows from the fact that $H - h$ is the value of k for which the former expression, regarded as a function of k alone, assumes its upper bound. Hence, bearing in mind the differentiability of power series (§ 42), the result of the preceding article is at once transformed into the following:

THEOREM. *The necessary and sufficient conditions that the series*

$$f(a) + hf'(a) + \tfrac{1}{2}h^2 f''(a) + \ldots$$

should converge for all values of h in the half-open interval

$$0 \leqslant h < b,$$

and have $f(a+h)$ for its sum, are that, (1) $f(a+h)$ and all its successive differential coefficients exist and are finite throughout the half-open interval, and (2) for each fixed positive value of H less than b,

$$\left| \frac{(H-h)^{n-p}}{(n-p)!} f^{(n)}(a+h) \right|$$

regarded as a function of the ensemble (h, n) should be bounded for all values of n such that $n - p \geqslant 0$, and all positive values of h less than or equal to H, p being zero, or any conveniently chosen positive or negative integer.

COR. *It is sufficient if besides (1) we know that*

$$\left| \frac{(b-h)^{n-p}}{(n-p)!} f^{(n)}(a+h) \right|$$

regarded as a function of (h, n) is bounded.

48. The following additional theorem at once follows:

THEOREM. *The necessary and sufficient conditions that the series*

$$f(a) + hf'(a) + \tfrac{1}{2}h^2 f''(a) + \ldots$$

should converge for all values of h in the closed interval $(0, b)$ and have $f(a+h)$ for its sum are that:

(1) $f(a+h)$ *and its differential coefficients should exist and be finite in the half-open interval* $0 \leqslant h < b$;

(2) *when* $h = b$, $f(a+h)$ *should be finite and continuous and the series should not oscillate*;

(3) *for each fixed positive value of* $H < b$,

$$\left| \frac{(H-h)^{n-p}}{(n-p)!} f^{(n)}(a+h) \right|$$

regarded as a function of the ensemble (h, n) *should be bounded for all values of* n *such that* $n - p \geqslant 0$ *and all positive values of* $h \leqslant H$.

That the condition (2) is necessary is evident, that conditions (1) and (3) are so also has just been proved. Further, if we suppose all three to hold, the series is necessarily convergent throughout the whole closed interval in question except possibly at b where it may diverge without oscillating, and therefore represents a bounded or unbounded continuous function throughout that closed interval. But its sum in the half-open interval is $f(a+h)$, by the preceding article, and $f(x)$ is a bounded continuous function in the whole closed interval. Hence the sum, when $h = b$, is $f(a+b)$.

49. The same for two or more variables. In the corresponding theorem for two or more variables, differentials take the place of differential coefficients, but we are no longer able to take both the steps analogous to those taken between § 46 and § 47. The extended theorem is as follows :

THEOREM. *If* $f(a+h, b+k)$ *and all its differentials exist throughout the partially open rectangle*

$$0 \leqslant h < R, \quad 0 \leqslant k < S,$$

in which case they are all necessarily continuous functions of the ensemble (h, k), *then the necessary and sufficient condition that the series*

$$f(a,\, b) + \left(h\frac{\partial}{\partial a} + k\frac{\partial}{\partial b} \right)f + \frac{1}{2!}\left(h\frac{\partial}{\partial a} + k\frac{\partial}{\partial b} \right)^{2}f + \dots$$

$$+ \frac{1}{n!}\left(h\frac{\partial}{\partial a} + k\frac{\partial}{\partial b} \right)^{n}f + \dots,$$

that is,

$$f(a,\, b) + hf_{a} + kf_{b} + \frac{1}{2!}\left(f_{aa}, f_{ab}, f_{bb} \right\rangle\!\langle h,\, k)^{2} + \dots$$

$$+ \frac{1}{n!}\left(\frac{\partial^{n}f}{\partial a^{n}},\, \dots\, \frac{\partial^{n}f}{\partial b^{n}} \right\rangle\!\langle h,\, k \right)^{n} + \dots,$$

should converge at every point of the partially open rectangle and have

$f(a + h, b + k)$ *for its sum is that for each pair of fixed positive values*
$H < R$ *and* $K < S$, *the expression*

$$\frac{1}{n!}\left(u\,\frac{\partial}{\partial h} + v\,\frac{\partial}{\partial k}\right)^n f(a + h, b + k)$$

(*where, after the expansion of the operator,* $(H - h)$, $(K - k)$ *are to be
written for* u, v *respectively), regarded as a function of the ensemble*
$(h, k; n)$, *should be bounded for all positive integral values of* n, *and all
positive values of* $h \leqslant H$, *and of* $k \leqslant K$, *such that* $k/h = K/H$.

Under these circumstances then the theorem states that

$$f(a + h,\; b + k) = f(a, b) + df + \frac{1}{2!}\,d^2f + \dots + \frac{1}{n!}\,d^nf + \dots,$$

where in using the notation df, d^2f, ... it is to be understood that no
restriction is put on the values of h and k other than that stated
above.

The statement for more than two variables is of course precisely
similar.

To prove the theorem we merely have to consider that it is necessary
and sufficient that Taylor's Theorem should hold on the portion of any
straight line through the corner (a, b) which lies in the rectangle.
Taking t as current coordinate on such a line, so that, by § 21,

$$\frac{d^rf}{dt^r} = \left(\cos\theta\,\frac{\partial}{\partial h} + \sin\theta\,\frac{\partial}{\partial k}\right)^r f,$$

where $\tan\theta = k/h = K/H$, and transforming the conditions in terms of t
got from § 47, the required result follows immediately.

APPENDIX I.

EXPLANATORY NOTES.

(a) § 2, p. 1. It is remarkable that, general as this definition of function is, the limits (§§ 3, 4) on the right must exactly coincide with those on the left except at a countable set of points in the case of a function of one variable [2],[3]. In the case of a function of more than one variable, the points at which there is a difference in the limits for approach by regions, if any, form a set of the first category [4].

(β) § 3, p. 2. The associated upper and lower limiting functions were first used by Baire [5], who, however, included the value at the point itself among the values to be considered in each of the defining intervals $d_1, d_2, \ldots$. The author introduced the usage in the text, and shewed that the inequality

$$\psi(x) \leqslant f(x) \leqslant \phi(x),$$

which with Baire's definition was true everywhere, holds at all but a countable set of points. Moreover the same is true if the limits are taken on one side only of each point, when there is only a single variable [2].

(γ) § 4, p. 3. It may happen that a limiting value is approached by means of values all greater (less) than the limit in question. We sometimes find it convenient to distinguish between the limits approached in these two ways, that is, we may attach to a limit a *sense* as well as a *magnitude* [6].

(δ) § 6, p. 6. For choosing proper sequences $x_1, x_2, \ldots, x_m, \ldots$ with x as limiting point, and $x_{m,1}, x_{m,2}, \ldots, x_{m,n}, \ldots$ with x_m as limiting point,

$$\text{upper limit } \underset{y=x}{\phi(y)} = \underset{m=\infty}{\text{Lt}} \phi(x_m) = \underset{m=\infty}{\text{Lt}} \underset{n=\infty}{\text{Lt}} f(x_{m,n}),$$

which is a repeated limit and therefore (§ 5) a double limit $(m, n) = (\infty, \infty)$ of $f(x_{m,n})$, that is one of the limits of $f(y)$, and therefore $\leqslant \phi(x)$.
$$\underset{y=x}{}$$

(ε) § 6, p. 7. For if f is upper semi-continuous and finite, its upper bound is finite, since f assumes this as a value. Hence, if f is not bounded, there must be a point P such that $f(P) < -k$, where k is any positive quantity, and therefore a whole interval throughout which $f(x) < -k$.

Either f is bounded in this interval, or, similarly, we get an interval inside the first throughout which $f(x) < -2k$, and so on. Thus either we arrive at an interval in which f is bounded, or we get a sequence of intervals, each inside the preceding, and having therefore at least one point Q internal to all of them, such that, therefore, $f(Q) < -nk$ for all integers k. This is however impossible, since $f(Q)$ is finite, which proves that there is an interval in which f is bounded. Q. E. D.

ζ) § 8, p. 8. Functions that are upper or lower semi-continuous on the right (left) alone have also been shewn to be pointwise discontinuous [10].

The limit of a sequence of functions which are at every point continuous on one side, not necessarily the same for each point, has been shewn to be a pointwise discontinuous function, and a suitable generalisation has been given for functions of more than one variable [9].

(η) § 9, p. 10. The upper (lower) derivates on the left and right only differ at a set of the first category, and this whether or no the derivates are bounded [11]. If the derivates are bounded they differ only at a set of zero content [12]. Regarded as functions—either upper derivate is upper semi-continuous, and either lower derivate lower semi-continuous, except at a set of the first category [11]. An upper (lower) derivate is the limit of a monotone descending (ascending) sequence of lower (upper) semi-continuous functions. All the derivates of $f(x)$ have the same upper and lower bounds in any interval [13], and these are the same whether the interval includes its end-points, or not, and are the same as those of the incrementary ratio ; moreover these bounds are unaltered if we omit any countable set of points from the interval [12], [14]. All the derivates have the same upper and lower limits on each side at every point and lie between them (inclusive) [15] ; hence if one derivate is continuous at a point so are they all and they are all equal, so that at that point there is a differential coefficient. Hence, in particular, if any derivate is known to be zero except at a countable set of points, and is therefore zero everywhere, it is the differential coefficient, and the function is a constant. At a point where all four derivates agree, they all have the same derivates [15].

Weierstrass proved by construction of an example that a continuous function may throughout an interval have no differential coefficient [16], [17]. In this case its derivates have the following properties [18] :

(1) There is necessarily a distinction of right and left in the values of the derivates at a set of points dense everywhere and of the first category ;

(2) The upper and lower bounds of the derivates at the points of this set are respectively $+\infty$ and $-\infty$;

(3) At the remaining points of the interval both the upper derivates are $+\infty$ and both the lower derivates $-\infty$, exception being made of at most another set of the first category.

(θ)   § 9, p. 11.   If the repeated limit

$$\underset{k=0\,h=0}{\text{Lt Llt}}\ (f(x+h+k)-f(x+h)-f(x+k)-f(x))/hk$$

exists, it follows at once from the properties of derivates that the single limit $\underset{h=0}{\text{Lt}}\ (f(x+h)-f(x))/h$ exists.   For each derivate of $(f(x+k)-f(x))/k$ has then an unique limit at $k=0$, and is therefore continuous and equal to all the other derivates.   Thus if we choose to define the second differential coefficient without assuming the existence of the first differential coefficient, as the repeated limit in the extended sense mentioned at the end of § 5, it follows of necessity that the first differential coefficient exists at the point x, though not in the neighbourhood.

(ι)   § 11, p. 13.   If we start with the Theorem of the Mean, the corresponding theory of Indeterminate Forms is a little different.   We then prove that, if as x approaches the value a, $f(x)$ and $F(x)$ have each the unique limit zero, then the limits of $f(x)/F(x)$ are, in sense as well as magnitude, limits of $f'(x)/F'(x)$, provided a is neither a limiting point of common infinities nor common zeros of f' and F'.   See the author's paper "On Indeterminate Forms" [6] where moreover it is shewn that the recently abandoned proof of the rule for ∞/∞, assuming that for $0/0$, only required slight modification to make it valid.

(κ)   § 13, p. 16.   It follows at once from the Expansion Theorem that, if the repeated limit (p. 11) which defines the second differential coefficient exists, then the double limit exists and is equal to it for all modes of approach of (h, k) to $(0, 0)$ which do not make k/h nor h/k zero.   In particular both

$$\{f(x+2h)-2f(x+h)+f(x)\}/h^2 \text{ and } \{f(x+h)+f(x-h)-2f(x)\}/h^2$$

have a definite limit if $f''(x)$ exists and its value is $2f''(x)$.   The converse is of course in general not true, but it is interesting to note that, if we know that throughout a whole interval this limit is continuous, or that the limit is a bounded function which is the differential coefficient of its (generalised or Lebesgue) integral, the converse holds also.

Notice that the Expansion Theorem will not replace the definition of the nth differential coefficient.   For example, it does not follow from the existence of an unique limit for $(f(x+h)-f(x)-hf'(x))/\tfrac{1}{2}h^2$ that $f''(x)$ exists.

(λ)   § 15, p. 18.   If we do not assume the existence of the differential coefficient at all, but only that there is no distinction of right and left between the derivates of $f(x)$, it will follow from this assumption that a differential coefficient does exist at points which are dense everywhere and of the second category, and, further, that the Theorem of the Mean holds for some point x of this set [15].

(μ)   § 15, pp. 19, 20.   In certain cases we may infer that the function is a constant without knowing a priori whether the differential coefficient exists.   Thus the theorem about the bounds of the derivates [7] enables us to assert that a function is a constant if one of the derivates at every point is known to be zero, except at a countable set, or, by a theorem of Lebesgue's[12], if the exceptional set is only of content zero, provided the bounds of the derivates are known to be finite.

(ν)   § 16, p. 21. The property of assuming all values between its upper and lower bounds belongs, as is pointed out in the present Tract, to the following functions :—Continuous functions, § 7, p. 7, the incrementary ratio, § 9, p. 10, the double incrementary ratio, § 16, p. 20, the differential coefficient, § 15, p. 18, the repeated differential coefficients, § 16, p. 20. The necessary and sufficient condition that a function which is the limit of a continuous function should assume all values between its upper and lower bounds in every interval is that the value of the function at each point is one of the limits of values in the neighbourhood of the point on the right and also on the left [19].

(π)   § 28, p. 35. This theorem has been stated in its most familiar form. A discussion of how far the limitations may be mitigated, with reference to Stolz's *Grundzüge der Differential- und Integralrechnung*, will be found in the author's paper[21].

(ρ)   § 41, p. 50. $\dfrac{\partial^2 f}{\partial x \partial y}$ and $\dfrac{\partial^2 f}{\partial y \partial x}$ have been shewn to have the same upper and lower bounds in every region throughout which they both exist, § 16, p. 21. From this we have an intuitive proof that if, at a point, one of the two is continuous, so is the other and the values are the same there.

APPENDIX II.

LIST OF BOOKS AND PAPERS QUOTED.

Supplementary Information and References to the Literature of the Subject may be found in the following List of Papers and Books quoted by Number in this Tract.

(1) "The Theory of Sets of Points," by W. H. Young and Grace Chisholm Young, 1906, Cambridge University Press.

(2) "On the Distinction of Right and Left at Points of Discontinuity," by W. H. Young, 1907, *Quart. J.* xxxix. pp. 67-83.

(3) "Sulle funzioni a piu valori costituite dai limiti d' una funzione d' una variabile reale a destra e a sinistra di un punto," id. 1908, *Rend. dei Lincei*, xvii. Ser. v. pp. 582-587.

(4) "On the Discontinuities of a Function of One or More Real Variables," id. 1909, *Proc. L.M.S.* Ser. 2, Vol. viii. pp. 117-124.

(5) "Sur les fonctions de variables réelles," by R. Baire, 1899, *Ann. di Mat.* Ser. iii. Vol. iii. pp. 1-123.

(6) "On Indeterminate Forms," by W. H. Young, 1909, *Proc. L.M.S.* Ser. 2, Vol. viii. pp. 39-76.

(7) "On Monotone Sequences of Continuous Functions," id. 1908, *Proc. Camb. Phil. Soc.* xiv. pp. 520-529.

(8) "A new Proof of a Theorem of Baire's," id. 1907, *Mess. Math.*, xxxvii. pp. 49-54. (Cp. (9).)

(9) "On Sequences of Asymmetrically Continuous Functions," id. 1909, *Quart. J.* xl. pp. 374-380.

(10) "Note on Left and Right-handed Semi-continuous Functions," id. 1908, *Quart. J.* xxxix. pp. 263-265.

(11) "Oscillating Successions of Continuous Functions," id. 1908, *Proc. L.M.S.* Ser. 2, Vol. vi. pp. 298-320.

(12) "Leçons sur l'intégration," by H. Lebesgue, 1904, Gauthier-Villars. See also various papers in the *Rend. dei Lincei*, where a small correction is made in the statement and proof on p. 122.

(13) "Fondamenti per la teorica delle funzioni di variabili reali," by U. Dini, 1878, German trans. by J. Lüroth and A. Schepp, 1892, Teubner.

(14) "On Term-by-term Integration of Oscillating Series," by W. H. Young, 1909, *Proc. L.M.S.* Ser. 2, Vol. viii. pp. 99-116.

(15) "On Derivates and the Theorem of the Mean," by W. H. Young and Grace Chisholm Young, 1908, *Quart. J.* XL. pp. 1–26; "Additional Note on Derivates and the Theorem of the Mean," id. pp. 144–5.

(16) "Zur Functionenlehre," by K. Weierstrass, 1880, in the *Abhandlungen aus der Functionenlehre*, pp. 97 seq.

(17) "The Theory of Functions of a Real Variable and the Theory of Fourier's Series," by E. W. Hobson, 1907, Cambridge University Press.

(18) "On the Derivates of Non-differentiable Functions," by W. H. Young, 1908, *Mess. Math.* XXXVIII. pp. 65–69.

(19) "A Theorem in the Theory of Functions of a Real Variable," id. 1907, *Rend. Palermo*, XXIV. pp. 1–6; "A Note on Functions of Two or More Variables which Assume all Values between their Upper and Lower Bounds," id. 1909, *Mess. Math.*, XXXIX. pp. 69—72.

(20) "On Differentials," id. 1908, *Proc. L.M.S.* Series 2, Vol. VII. pp. 157–180.

(21) "Grundzüge der Differential- und Integralrechnung," by O. Stolz, 1893, Teubner. "Note on a Remainder Form of Taylor's Theorem," by W. H. Young, 1908, *Quart. J.* XL. pp. 146–153.

(22) "Implicit Functions and their Differentials," id. 1909, *Proc. L.M.S.* Ser. 2, Vol. VII. pp. 398–421.

(23) "On the Conditions for the Reversibility of the Order of Partial Differentiation," id. 1909, *Proc. R.S. Edinburgh*, Vol. XXIX. pp. 136–167.

(24) "Ueber die notwendigen und hinreichenden Bedingungen des Taylor'-schen Lehrsatzes für Funktionen einer reellen Variabeln," by A. Pringsheim, *Math. Ann.* XLIV.; "On Taylor's Theorem," by W. H. Young, 1908, *Quart. J.* XL. pp. 157–167; cp. also (17).

(25) "Sopra le funzioni che hanno derivata in ogni punto," by Beppo Levi, 1906, *Rend. dei Lincei*, XV. Serie 5, pp. 410–415.

APPENDIX III.

ON THE EXPRESSIONS AND RESULTS BORROWED FROM THE THEORY OF SETS OF POINTS USED IN THIS TRACT.

(*" The Theory of Sets of Points," by W. H. Young and Grace Chisholm Young, Cambridge University Press*, 1906, *is cited below as " Th. S."; the Roman numerals refer to the chapters. The reference at the beginning of each of the following notes is to the present Tract.*)

§ 3, p. 2, lines 7 seq. See Th. S. III. p. 17, § 10.

This process of taking intervals one inside the other requires to be grasped for the proper understanding of all pure mathematics; it is intimately connected with the very concept of an irrational number. When we say "take" a set of intervals $d_1, d_2, \ldots d_n, \ldots$, we mean: imagine a law given, involving a variable integer n, such that inserting for n any chosen integer, we get an interval determined.

For instance, on the straight line we might have the law

$$d_n \equiv (a - 2^{-n} \leqslant x \leqslant a + 2^{-n}),$$

or in the plane the "interval" d_n might be a square of side 2^{-n} with the point a, that is (a_1, a_2), as centre. In these instances each interval d_n lies inside the preceding interval d_{n-1}.

Whatever integer be chosen, the corresponding interval is to be considered, and no other intervals are to be taken into consideration. We may then properly speak of "all" the intervals, just as we may say "all" the integers, although we have no process by which we can present more than a selection of them *simultaneously* to the mind. There are however facts which can be stated about *all* the intervals, independent of such simultaneous presentation. In particular if, as in the examples quoted, (in which case the *set* is called *a sequence*), the intervals become smaller than any assigned magnitude as n increases, there will be one and only one point belonging to all the intervals, and this point will be an internal point of every interval if, as in the examples given, the end-points of each interval are different from those of its predecessor. In fact, not only is a a possible value of x when $a - 2^{-n} < x < a + 2^{-n}$, whatever integer n may be, but also, taking any other number b, we can find an integer m such that 2^{-m} is less than the difference of a and b, in which case b will not be a possible value of x when

$$a - 2^{-m} \leqslant x \leqslant a + 2^{-m},$$

so that the point b is not a point of d_m, and therefore not of d_n when $n \geqslant m$. Thus the point a is internal to all the intervals and is the only common point of the intervals.

An "intermediate interval" (§ 3, line 17) means an interval contained in d_n and containing d_{n+1}, for some value of n.

§ 3, p. 2, line 29. See Th. S. III. § 10, pp. 16–18 ; VIII. §§ 103, 104.

A " set " of points may consist of a finite number of points or of an infinite number of points determined by some consistent law involving a variable element. In the latter case the same remarks apply to the use of the word "all" as in the preceding paragraph. The variable element may be, but need not be, an integer ; it is very commonly a continuous variable x. Sometimes the law only involves the variable element implicitly, as for instance in the case of the set of numbers less than unity whose expressions in the ternary scale involve the figures 0 and 2 only. (See Th. S. Ex. 2, pp. 20 and 48.) Here the variable element does not formally appear but may be shewn to be *essentially* a continuous variable. A set whose variable element is essentially an integer is said to be *countably infinite*; one whose variable element is essentially a continuous variable is said to *have the potency c*. (See Th. S. IV. §§ 17, 18, 22, 23.)

A point L is said to be *a limiting point of a given set* if in every neighbourhood of the point L there is a point of the set. Here, as elsewhere, a *neighbourhood* or *open neighbourhood* of a point L consists of the points *other than L* in an interval containing L as internal or end-point while *a closed neighbourhood* includes also the point L. Thus the point a is the sole limiting point of the set consisting of the end-points of the above intervals d_n, i.e. the points $a \pm 2^{-n}$.

Any point of the closed interval (0, 1) is a limiting point of the set of rational points $x = m/n$, where n is any integer and m any smaller integer prime to n. Any point of the plane is a limiting point of the set of rational points $\left(\dfrac{m}{n}, \dfrac{p}{q}\right)$, where m, n, p and q are any integers.

§ 4, p. 3, lines 14–20. See Th. S. III. § 10, p. 19 ; §§ 12–14, pp. 23–29.

The *first derived set* of a set G is the set of all the limiting points of G.

A *closed set* is a set which contains all its limiting points ; in particular it contains those points which bound the set above and below (cp. p. 3, lines 16 and 26, and p. 38, line 18).

CANTOR'S THEOREM OF DEDUCTION. "*If G_1, G_2, ... be a series of closed sets of points, each of which is contained in the preceding set, then there is at least one point common to all the sets, and all such common points form a closed set.*"

§ 4, p. 3, line 30. See Th. S. III. p. 18, Theorem 2 and VIII. p. 175, Theorem 1.

A *sequence of points* means a set having one and only one limiting point.

72   FUNDAMENTAL THEOREMS OF DIFFERENTIAL CALCULUS

§ 7, p. 8, line 6.   It is clearly sufficient to prove that *an interval* (*a, b*) *on a straight line cannot be divided into two closed sets G and H without common points.*   Suppose the contrary is the case.   There are points of one of the sets, say of *G*, as near as we please to *a*, so that *a*, being a limiting point of the set *G*, is a point of *G*, that set being closed.   Hence *a* is not a point of the other set *H*, and therefore there are intervals with *a* as left-hand end-point, containing no point of *H*.   Let *A P* be that interval whose length is the upper bound of the lengths of those intervals.

Then every internal point of the interval *A P* belongs to the set *G*, therefore the same is true of *P*, since *G* is closed.   Therefore, as in discussing the point *a*, there is an interval *A Q* larger than *A P* containing no point of *H*. But this is not true ; therefore etc.   Q. E. D.

§ 8, p. 8, line 38, and p. 9, line 16.   Th. S. p. 21.   A set is said to be *dense everywhere* in a given interval when there is a point of the set in every interval inside the given interval.   A set is said to be *dense nowhere* in a given interval, if it is not dense everywhere in any interval inside the given interval; in this case there is in every interval inside the given interval an interval containing no point of the set.

The rational points are dense everywhere and so are the irrational points ; the ternary fractions involving only the figures 0 and 2 are dense nowhere in the interval (0, 1).

A closed set dense everywhere consists therefore of all the points of an interval.   A closed set dense nowhere consists of the end-points and external points of a set of non-overlapping intervals dense everywhere.

The process used in the text of proving that there is a point which does not belong to any of the sets of points at which $\phi(x) - f(x) \geqslant k$, for any positive value of k, is one that is constantly used, and has led to the use of Baire's term *set of the first category*[6] for the set of those points which belong to at least one set (and therefore to all subsequent sets) of a sequence of sets $G_1, G_2, \ldots G_n, \ldots$ each containing its predecessor and dense nowhere.   As in the text, where the set G_n may be taken to consist of all the points at which $\phi(x) - f(x) \geqslant k\,2^{-n}$, we get for each value of n an interval d_n containing no point of G_n and lying in d_{n-1}.   Hence there is a point P internal to all these intervals, and which accordingly belongs to none of the sets G_n.   This shews that *a continuum is not a set of the first category.*

Printed in the United States
By Bookmasters